HANDYMAN
CLUB OF AMERICA
P9-DDF-656

Advanced Wiring & Plumbing Projects

Complete Handyman's Library™
Handyman Club of America
Minneapolis, Minnesota

Published in 1997 by
Handyman Club of America
12301 Whitewater Drive
Minnetonka, Minnesota 55343

Published by arrangement with Cowles Creative Publishing, Inc.
ISBN 0-86573-658-8

Printed on American paper by
R. R. Donnelley & Sons Co.
99 98 97 / 5 4 3 2 1

CREDITS:
Created by: The Editors of Cowles Creative Publishing and the staff of the Handyman Club of America in cooperation with Black & Decker. BLACK&DECKER is a trademark of Black & Decker (US), Incorporated and is used under license.

Handyman Club of America:
Vice President, Products & Business Development: Mike Vail
Book Products Development Manager: Mark Johanson
Book Marketing Coordinator: Jay McNaughton

NOTICE TO READERS

This book provides useful instructions, but we cannot anticipate all of your working conditions or the characteristics of your materials and tools. For safety, you should use caution, care, and good judgment when following the procedures described in this book. Consider your own skill level and the instructions and safety precautions associated with the various tools and materials shown. Neither the publisher nor Black & Decker® can assume responsibility for any damage to property or injury to persons as a result of misuse of the information provided.

The instructions in this book conform to "The Uniform Plumbing Code," "The National Electrical Code Reference Book," and "The Uniform Building Code" current at the time of its original publication. Consult your local Building Department for information on building permits, codes, and other laws as they apply to your project.

Contents

AUTOMATIC LOCK

Advanced Wiring Projects

Installing new wiring in your home remodeling projects is quite simple when you know the basic principles and requirements. You also will save a great deal of money by doing the work yourself—and you will know exactly how your house is wired and that the work has been done well. This section of *Advanced Wiring & Plumbing Projects* shows the essential information for doing this work, from planning your wiring project through its final inspection. If you have done basic electrical repairs, you can easily accomplish your own major wiring project.

First you will learn about *Planning a Wiring Project*. You will find out which Electrical Code requirements apply to your project, as well as how to work with your local electrical inspector. You will see how to evaluate your existing electrical capacity and power usage, and determine the needs of the circuits you are adding. You will also learn how to draw a wiring diagram you will use to obtain a permit and successfully perform the work.

Then three major wiring projects demonstrate necessary methods and techniques. *Wiring a Room Addition* presents information that applies to any addition, whether it is an attic or basement renovation, or a new room added on to your existing house. In addition to basic receptacles and light fixtures, you will learn how to install circuits and fixtures for a ceiling fan, permanently wired smoke alarm, bathroom vent fan, computer receptacle, air-conditioning receptacle, electric heaters, telephone outlets, and cable television jacks.

In *Wiring a Remodeled Kitchen* you will find how to wire a kitchen or any other room that uses a lot of power and contains many specialty circuits. You will learn how to install circuits and fixtures for recessed lights, under-cabinet task lights, and a ceiling light controlled by three-way switches. You will also learn how to install circuits and receptacles for a range, microwave, dishwasher, and food disposer. Methods for installing two small-appliance circuits are also shown.

Finally, *Installing Outdoor Wiring* shows you how to get power to any outdoor location. You will learn how to install circuits and fixtures for decorative lights, weatherproof switches, GFCI-protected receptacles, and a manual override switch that lets you control a motion-sensor light fixture from inside the house.

Your remodeling projects and their wiring needs will differ from the projects shown in *Advanced Wiring & Plumbing Projects*. But the concepts and methods shown apply to any wiring projects containing any combination of circuits. While the information presented in this section is based on current National Electrical Code regulations, always check with your local electrical inspector to make certain your project meets local requirements (which take precedence over those in the National Code).

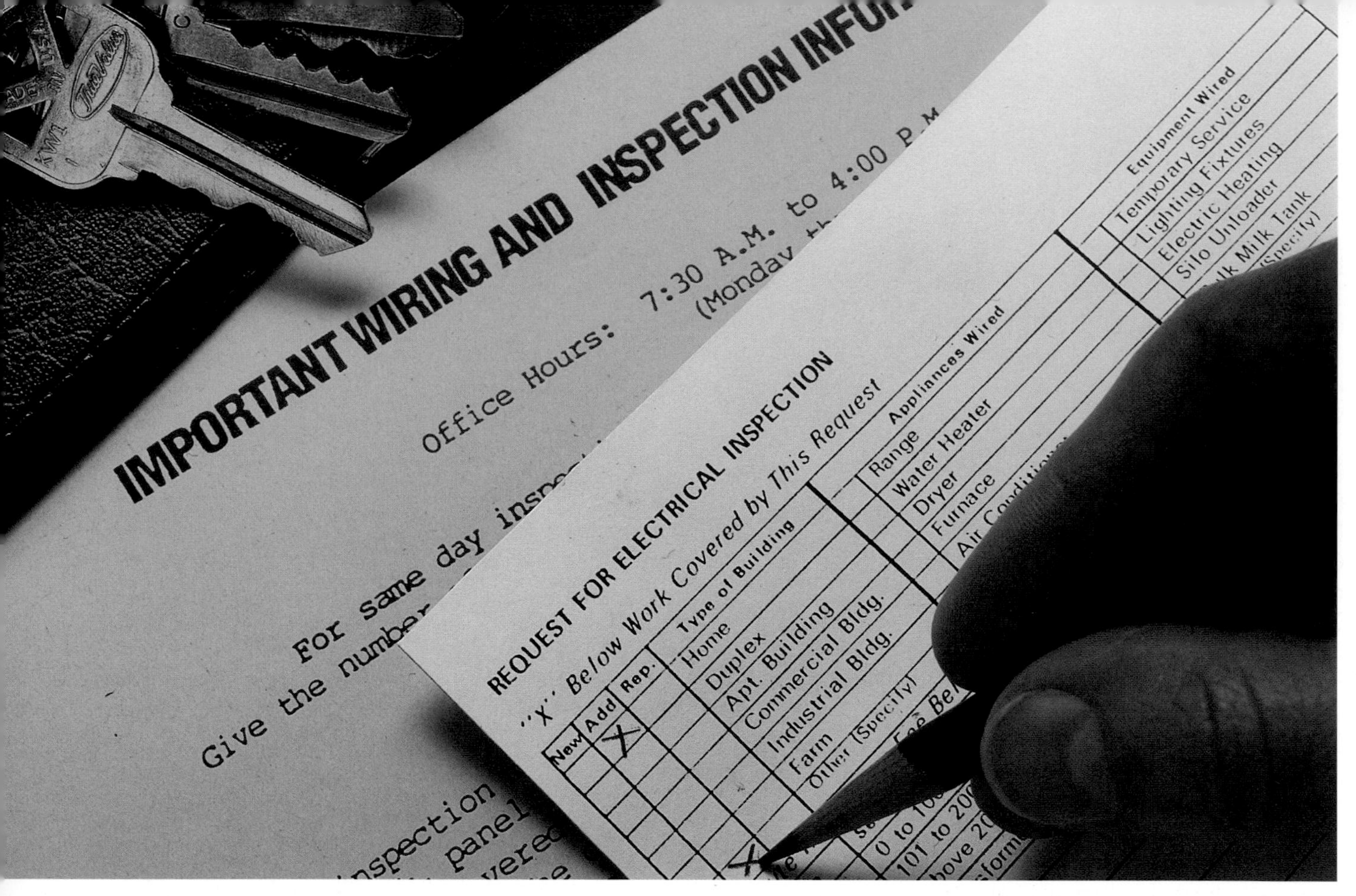

Remember that your work must be reviewed by your local electrical inspector. Follow the inspector's guidelines, and your careful planning and work will result in quality workmanship, and your project will pass inspection easily.

Planning a Wiring Project

Careful planning of a wiring project ensures that you will have plenty of power for present and future needs. Whether you are adding circuits in a room addition, wiring a remodeled kitchen, or adding an outdoor circuit, consider all possible ways the space might be used, and plan for enough electrical service to meet peak needs.

For example, remember that a room's use may change. A spare bedroom requiring only a single 15-amp circuit might need two 20-amp circuits if it becomes a home office or recreation space. When wiring a remodeled kitchen, it is a good idea to install circuits for an electric oven and countertop range, even if you do not have these appliances. It will be much easier to convert from gas to electric in the future.

A large wiring project adds a considerable load to your main electrical service. It may require upgrading the main service (especially if it is rated under 100-amps) or installing a subpanel (if the main service panel is too full to hold new circuit breakers). These are jobs for a licensed electrician, but are well worth the investment.

Planning a wiring project requires that you:

- **Examine your main service**
- **Learn about codes**
- **Prepare for inspections**
- **Evaluate electrical loads**
- **Draw a wiring diagram and get a permit**

Your inspector needs to see an accurate wiring diagram that shows the layout of the circuits you are installing and a materials list before he will issue a work permit for your project. This wiring plan also helps you organize your work.

Draw a scaled diagram of the space you will be wiring, showing walls, doors, windows, plumbing pipes and fixtures, and heating and cooling ducts. Mark the location of all the devices you will install with the symbols shown opposite. Then draw the cable runs (indicating cable size and type) between devices. You can see ex amples of project wiring diagrams on pages 16 to 17 and 38 to 39.

Before you can draw this wiring diagram, you must evaluate the existing electrical load in your home and the additional load created by your project.

To evaluate electrical load:
1. Multiply the square footage of all living areas times 3 watts.
2. Add 1500 watts for each kitchen small appliance circuit and for the laundry circuit.
3. Add ratings for permanent electrical appliances, including range, food disposer, dishwasher, freezer, water heater, and clothes dryer.
4. Determine the larger figure of total heating or total cooling wattages.
5. Multiply each outdoor receptacle (including those in garage) times 180 watts.
6. Add wattage ratings for outdoor fixtures (including those in garage).
7. Total these wattages to determine gross load. Figure the first 10,000 watts of the gross load at 100%. Figure the remaining gross load watts at 40%.
8. Add these two figures together to estimate the true electrical load.
9. Convert this estimate to amps by dividing by 230.
10. Compare this estimated amp load to the amp rating of your home's electrical service.

On the following pages you will see how to examine your main service, learn about codes, and prepare for the inspection.

Appliance Load Evaluation Chart

Item	**Range** (in watts)
Disposer	500-900
Dishwasher	1000-1500
Electric Range	3000-12,000
Microwave	500-800
Freezer	500-1000
Electric Clothes Dryer	4500-5500
Furnace	1000-20,000
Central Air	2300-5500
Window Air	500-2000
Baseboard Heater (per linear foot)	180-250
Electric Water Heater	3500-4500

Find wattage ratings for permanent appliances by reading the manufacturer's nameplate. If it gives the rating in kilowatts, find watts by multiplying kilowatts times 1000. If it lists only amps, find watts by multiplying amps times the voltage—either 120 or 240 volts, depending on the appliance. Freezers are permanent appliances requiring dedicated 15-amp, 120-volt circuits. Combination refrigerator-freezers rated at 1000 watts or less can be plugged into a small appliance circuit.

Electrical Symbol Key

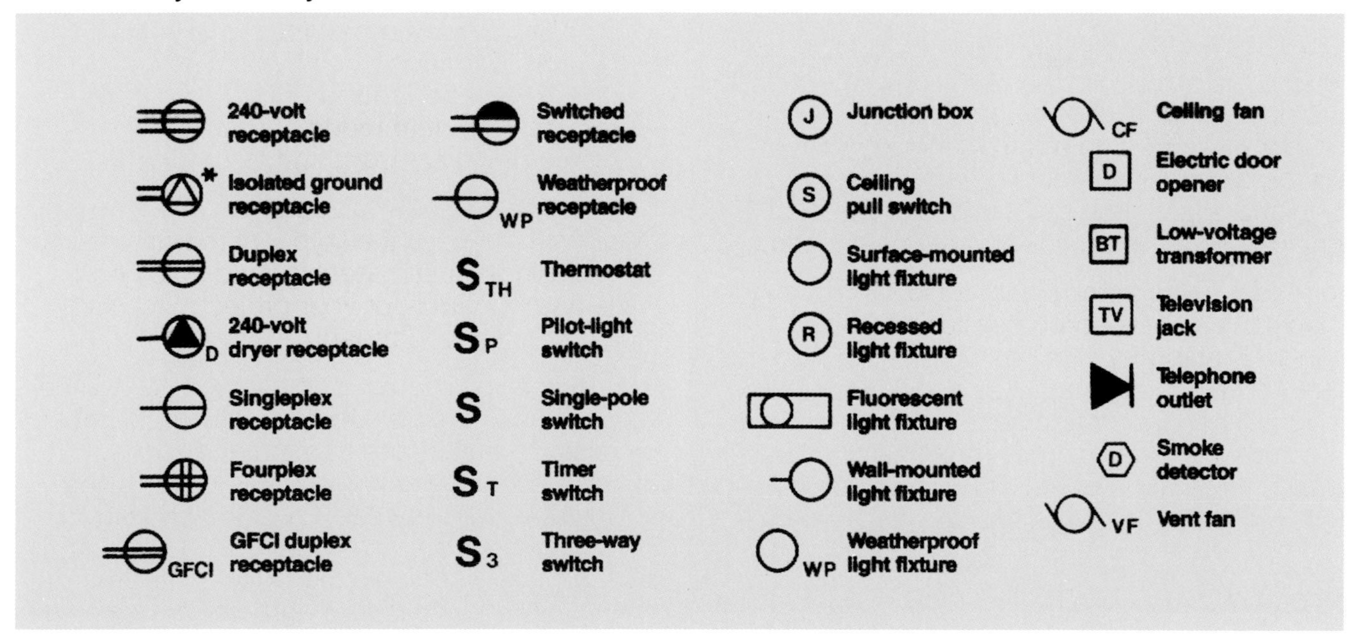

Use these standard electrical symbols when drawing a wiring diagram. Be sure to copy this key and attach it to your wiring plan for the inspector's convenience.

Examine Your Main Service

The first step in planning a new wiring project is to look in your main circuit breaker panel and find the size of the service by reading the amperage rating on the main circuit breaker. As you plan new circuits and evaluate electrical loads, knowing the size of the main service helps you determine if you need a service upgrade.

Also look for open circuit breaker slots in the panel. The number of open slots will determine if you need to add a circuit breaker subpanel.

Find the service size by opening the main service panel and reading the amp rating printed on the main circuit breaker. In most cases, 100-amp service provides enough power to handle the added loads of projects like the ones shown in this book. A service rated for 60 amps or less may need to be upgraded.

Older service panels use fuses instead of circuit breakers. Have an electrician replace this type of panel with a circuit breaker panel that provides enough power and enough open breaker slots for the new circuits you are planning.

Look for open circuit breaker slots in the main circuit breaker panel, or in a circuit breaker subpanel, if your home already has one. You will need one open slot for each 120-volt circuit you plan to install, and two slots for each 240-volt circuit. If your main circuit breaker panel has no open breaker slots, install a subpanel to provide room for connecting new circuits.

Learn About Codes

To ensure public safety, your community requires that you get a permit to install new wiring and have the completed work reviewed by an appointed inspector. Electrical inspectors use the National Electrical Code (NEC) as the primary authority for evaluating wiring, but they also follow the local Building Code and Electrical Code standards.

As you begin planning new circuits, call or visit your local electrical inspector and discuss the project with him. The inspector can tell you which of the national and local Code requirements apply to your job, and may give you a packet of information summarizing these regulations. Later, when you apply to the inspector for a work permit, he will expect you to understand the local guidelines as well as a few basic National Electrical Code requirements.

The National Electrical Code is a set of standards that provides minimum safety requirements for wiring installations. It is revised every three years. For more information, you can find copies of the current NEC, as well as a number of excellent handbooks based on the NEC, at libraries and bookstores.

In addition to being the final authority on Code requirements, inspectors are electrical professionals with years of experience. Although they have busy schedules, most inspectors are happy to answer questions and help you design well-planned circuits.

Basic Electrical Code Requirements

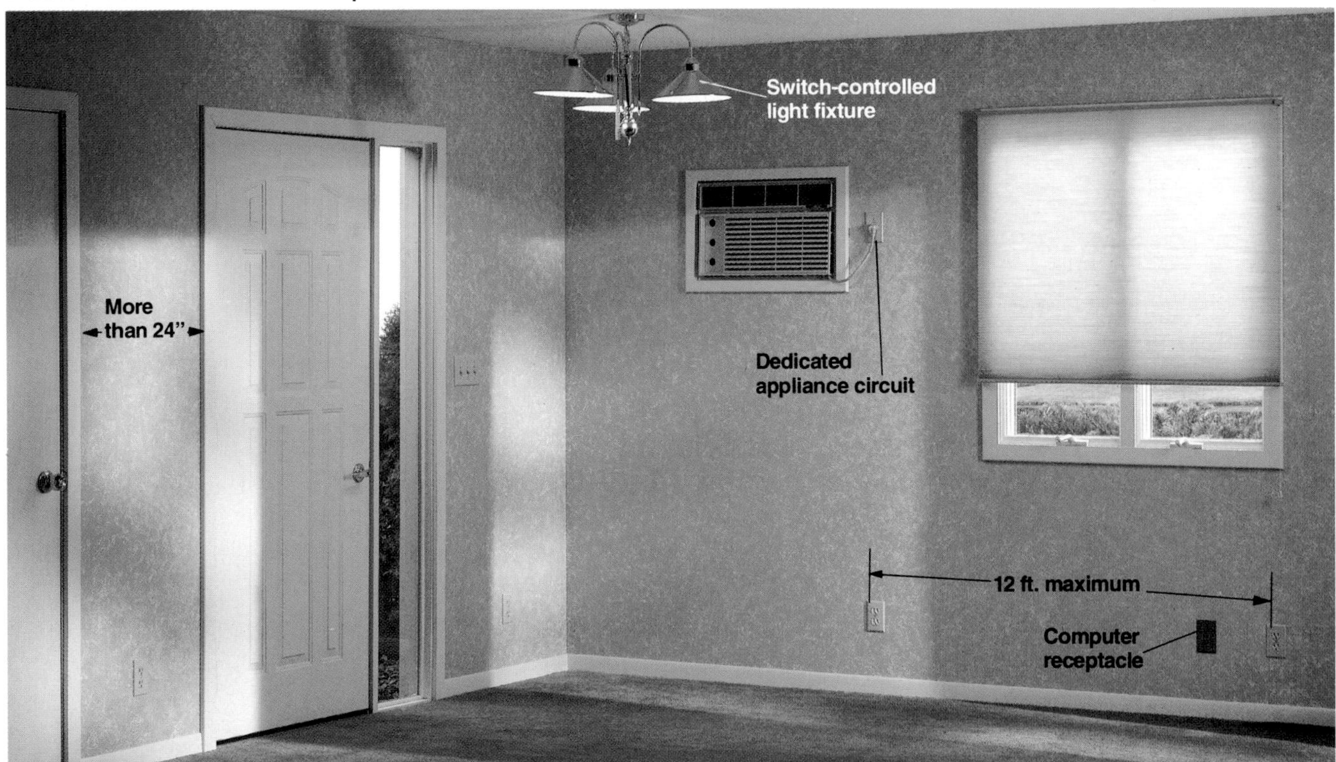

Electrical Code requirements for living areas: Living areas need at least one 15-amp or 20-amp basic lighting/receptacle circuit for each 600 square feet of living space, and should have a "dedicated" circuit for each type of permanent appliance, like an air conditioner, computer, or a group of baseboard heaters. Receptacles on basic lighting/receptacle circuits should be spaced no more than 12 ft. apart. Many electricians and electrical inspectors recommend even closer spacing. Any wall more than 24" wide also needs a receptacle. Evey room should have a wall switch at point of entry to control either a ceiling light or plug-in lamp. Kitchens and bathrooms must have a ceiling-mounted light fixture.

Prepare for Inspections

The electrical inspector who issues the work permit for your wiring project will also visit your home to review the work. Make sure to allow time for these inspections as you plan the project. For most projects, an inspector makes two visits.

The first inspection, called the "rough-in," is done after the cables are run between the boxes, but before the insulation, wallboard, switches, and fixtures are installed. The second inspection, called the "final," is done after the walls and ceilings are finished and all electrical connections are made.

When preparing for the rough-in inspection, make sure the area is neat. Sweep up sawdust and clean up any pieces of scrap wire or cable insulation. Before inspecting the boxes and cables, the inspector will check to make sure all plumbing and other mechanical work is completed. Some electrical inspectors will ask to see your building and plumbing permits.

At the final inspection, the inspector checks random boxes to make sure the wire connections are correct. If he sees good workmanship at the selected boxes, the inspection will be over quickly. However, if he spots a problem, the inspector may choose to inspect every connection.

Inspectors have busy schedules, so it is a good idea to arrange for an inspection several days or weeks in advance. In addition to basic compliance with Code, the inspector wants your work to meet his own standards for workmanship. When you apply for a work permit, make sure you understand what the inspector will look for during inspections.

You cannot put new circuits into use legally until the inspector approves them at the final inspection. Because the inspector is responsible for the safety of all wiring installations, his approval means that your work meets professional standards. If you have planned carefully and done your work well, electrical inspections are routine visits that give you confidence in your own skills.

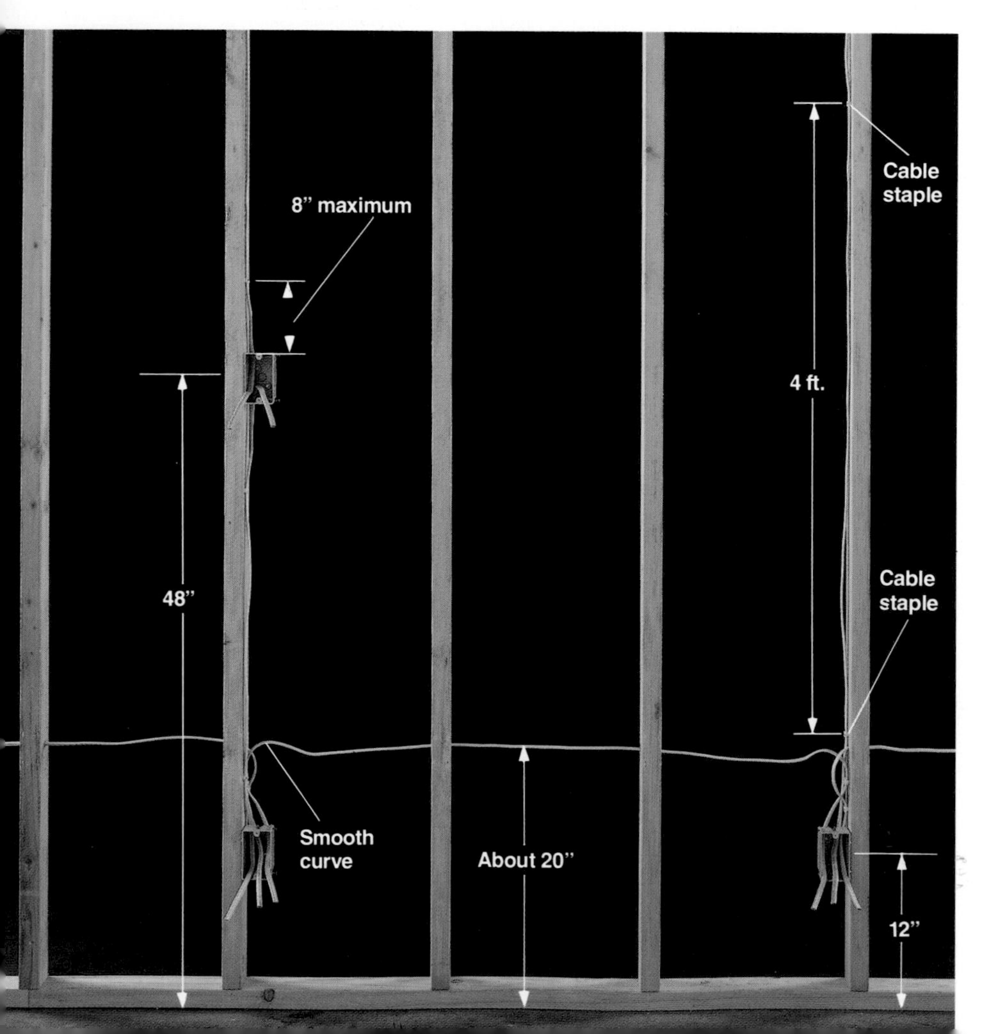

Inspectors measure to see that electrical boxes are mounted at consistent heights. Measured from the center of the boxes, receptacles in living areas typically are located 12" above the finished floor, and switches at 48". For special circumstances, inspectors allow you to alter these measurements. For example, you can install switches at 36" above the floor in a child's bedroom, or set receptacles at 24" to make them more convenient for someone in a wheelchair.

Your inspector will check cables to see that they are anchored by cable staples driven within 8" of each box, and every 4 ft. thereafter when they run along studs. When bending cables, form the wire in a smooth curve. Do not crimp cables sharply or install them diagonally between framing members. Some inspectors specify that cables running between receptacle boxes should be about 20" above the floor.

What Inspectors Look For

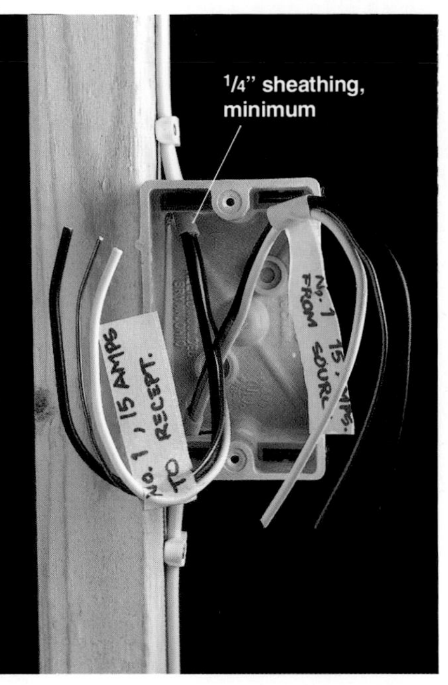

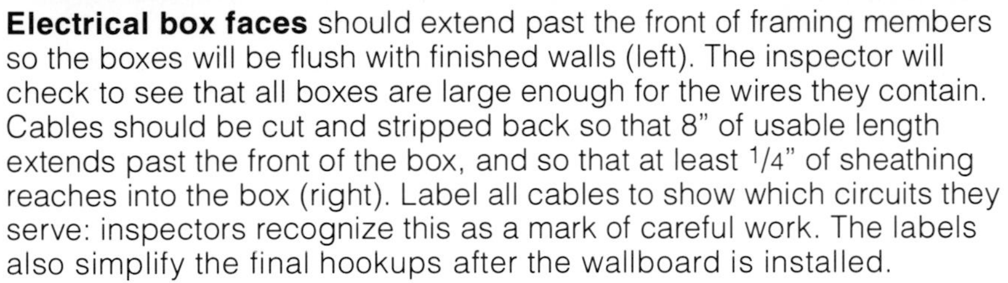

Electrical box faces should extend past the front of framing members so the boxes will be flush with finished walls (left). The inspector will check to see that all boxes are large enough for the wires they contain. Cables should be cut and stripped back so that 8" of usable length extends past the front of the box, and so that at least 1/4" of sheathing reaches into the box (right). Label all cables to show which circuits they serve: inspectors recognize this as a mark of careful work. The labels also simplify the final hookups after the wallboard is installed.

Install an isolated-ground circuit and receptacle if recommended by your inspector. An isolated-ground circuit protects sensitive electronic equipment, like a computer, against tiny current fluctuations. Computers also should be protected by a standard surge protector.

Heating & Air Conditioning Chart (compiled from manufacturers' literature)

Room Addition Living Area	Recommended Total Heating Rating	Recommended Circuit Size	Recommended Air-Conditioner Rating	Recommended Circuit Size
100 sq. feet	900 watts	15-amp (240 volts)	5,000 BTU	15-amp (120 volts)
150 sq. feet	1,350 watts		6,000 BTU	
200 sq. feet	1,800 watts		7,000 BTU	
300 sq. feet	2,700 watts		9,000 BTU	
400 sq. feet	3,600 watts	20-amp (240 volts)	10,500 BTU	
500 sq. feet	4,500 watts	30-amp (240 volts)	11,500 BTU	20-amp (120 volts)
800 sq. feet	7,200 watts	two 20-amp	17,000 BTU	15-amp (240 volts)
1,000 sq. feet	9,000 watts	two 30-amp	21,000 BTU	20-amp (240 volts)

Electric heating and air-conditioning for a new room addition will be checked by the inspector. Determine your heating and air-conditioning needs by finding the total area of the living space. Choose electric heating units with a combined wattage rating close to the chart recommendation above. Choose an air conditioner with a BTU rating close to the chart recommendation for your room size. NOTE: These recommendations are for homes in moderately cool climates, sometimes referred to as "Zone 4" regions. Cities in Zone 4 include Denver, Chicago, and Boston. In more severe climates, check with your electrical inspector or energy agency to learn how to find heating and air-conditioning needs.

Wiring a Room Addition

This chapter shows how to wire an unfinished attic space that is being converted to a combination bedroom, bathroom, and study. In addition to basic receptacles and light fixtures, you will learn how to install a ceiling fan, permanently wired smoke alarm, bathroom vent fan, computer receptacle, air-conditioning receptacle, electric heaters, telephone outlets, and cable television jacks. Use this chapter as a guide for installing your own circuits. Our room addition

Choose the Fixtures You Need

A: Computer receptacle (circuit #2) is connected to a 120-volt isolated-ground circuit. It protects sensitive computer equipment from power surges. See page 28.

B: Air-conditioner receptacle (circuit #3) supplies power for a 240-volt window air conditioner. See page 28. Some air conditioners require 120-volt receptacles.

C: Circuit breaker subpanel controls all attic circuits and fixtures, and is connected to the main service panel. For a more finished appearance, cover the subpanel with a removable bulletin board or picture.

D: Thermostat (circuit #5) controls 240-volt baseboard heaters in the bedroom and study areas. See page 32.

E: Fully wired bathroom (circuit #1) includes vent fan with timer switch, GFCI receptacle, vanity light, and single-pole switch. See pages 26 to 27. The bathroom also has a 240-volt blower-heater controlled by a built-in thermostat (circuit #5, page 32).

F: Closet light fixture (circuit #1) makes a closet more convenient. See page 27.

G: Smoke alarm (circuit #4) is an essential safety feature of any sleeping area. See page 31.

features a circuit breaker subpanel that has been installed in the attic to provide power for five new electrical circuits. Turn the page to see how these circuits look inside the walls.

Three Steps for Wiring a Room Addition

1. Plan the Circuits (pages 18 to 19).
2. Install Boxes & Cables (pages 20 to 25).
3. Make Final Connections (pages 26 to 33).

H: Double-gang switch box (circuit #4) contains a three-way switch that controls stairway light fixture and single-pole switch that controls a switched receptacle in the bedroom area. See page 29.

I: Fan switches (circuit #4) include a speed control for ceiling fan motor and dimmer control for the fan light fixture. See page 29.

J: Ceiling fan (circuit #4) helps reduce summer cooling costs and winter heating bills. See page 30.

K: Stairway light (circuit #4) illuminates the stairway. It is controlled by three-way switches at the top and bottom of the stairway. See page 31.

L: Cable television jack completes the bedroom entertainment corner. See page 24.

M: Telephone outlet is a convenient addition to the bedroom area. See page 25.

N: Switched receptacle (circuit #4) lets you turn a table lamp on from a switch at the stairway. See page 30.

O: Receptacles (circuit #4) spaced at regular intervals allow you to plug in lamps and small appliances wherever needed. See page 30.

P: Baseboard heaters (circuit #5) connected to a 240-volt circuit provide safe, effective heating. See page 33.

Wiring a Room Addition: Construction View

The room addition wiring project on the following pages includes the installation of five new electrical circuits: two 120-volt basic lighting/ receptacle circuits, a dedicated 120-volt circuit with a special "isolated" grounding connection for a home computer, and two 240-volt circuits for air conditioning and heaters. The photo below shows how these circuits look behind the finished walls of a room addition.

Learn How to Install These Circuits & Cables

#1: Bathroom circuit. This 15-amp, 120-volt circuit supplies power to bathroom fixtures and to fixtures in the adjacent closet. All general-use receptacles in a bathroom must be protected by a GFCI.

#2: Computer circuit. A 15-amp, 120-volt dedicated circuit with an extra isolated grounding wire that protects computer equipment.

Circuit breaker subpanel receives power through a 10-gauge, three-wire feeder cable connected to a 30-amp, 240-volt circuit breaker at the main circuit breaker panel. Larger room additions may require a 40-amp or a 50-amp "feeder" circuit breaker.

#3: Air-conditioner circuit. A 20-amp, 240-volt dedicated circuit. In cooler climates, or in a smaller room, you may need an air conditioner and circuit rated for only 120 volts (page 11).

Wiring a room addition is a complex project that is made simple by careful planning and a step-by-step approach. Divide the project into convenient steps, and complete the work for each step before moving on to the next.

Tools You Will Need:

Marker, tape measure, calculator, screwdriver, hammer, crescent wrench, jig saw or reciprocating saw, caulk gun, power drill with 5/8" spade bit, cable ripper, combination tool, wallboard saw, needlenose pliers.

■ **#4: Basic lighting/ receptacle circuit.** This 15-amp, 120-volt circuit supplies power to most of the fixtures in the bedroom and study areas.

■ **#5: Heater circuit.** This 20-amp, 240-volt circuit supplies power to the bathroom blower-heater and to the baseboard heaters. Depending on the size of your room and the wattage rating of the baseboard heaters, you may need a 30-amp, 240-volt heating circuit.

Telephone outlet is wired with 22-gauge four-wire phone cable. If your home phone system has two or more separate lines, you may need to run a cable with eight wires, commonly called "four-pair" cable.

Cable television jack is wired with coaxial cable running from an existing television junction in the utility area.

Wiring a Room Addition: Diagram View

This diagram view shows the layout of five circuits and the location of the switches, receptacles, lights, and other fixtures in the attic room addition featured in this chapter. The size and number of circuits, and the list of required materials, are based on the needs of this

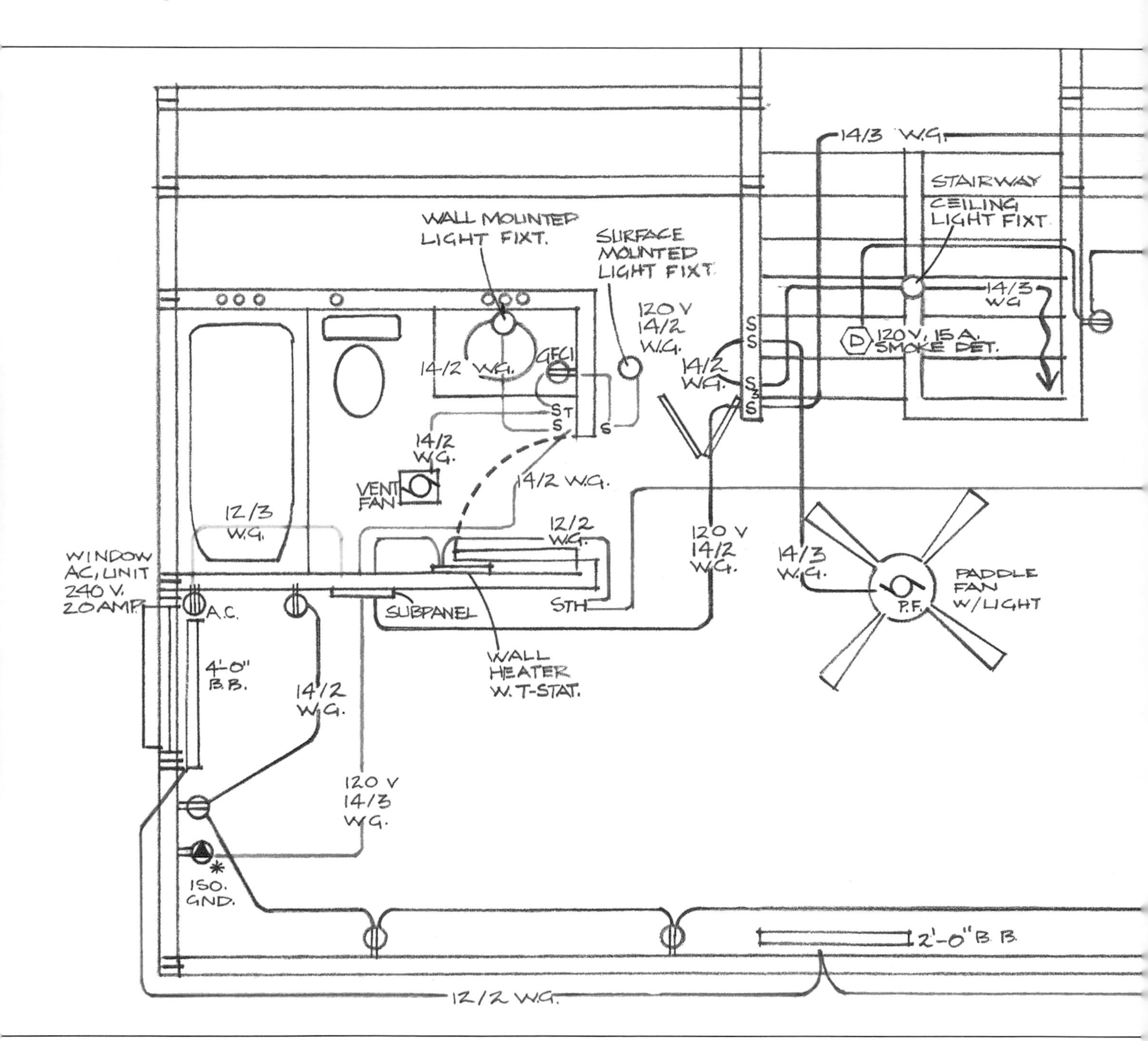

Circuit #1: A 15-amp, 120-volt circuit serving the bathroom and closet area. 14/2 NM cable, double-gang box, timer switch, single-pole switch, 4" × 4" box with single-gang adapter plate, GFCI receptacle, 2 plastic light fixture boxes, vanity light fixture, closet light fixture, 15-amp single-pole circuit breaker.

Circuit #2: A 15-amp, 120-volt computer circuit. 14/3 NM cable, single-gang box, 15-amp isolated-ground receptacle, 15-amp single-pole circuit breaker.

400-sq. ft. space. No two room additions are alike, so you will need to create a separate wiring diagram to serve as a guide for your own wiring project.

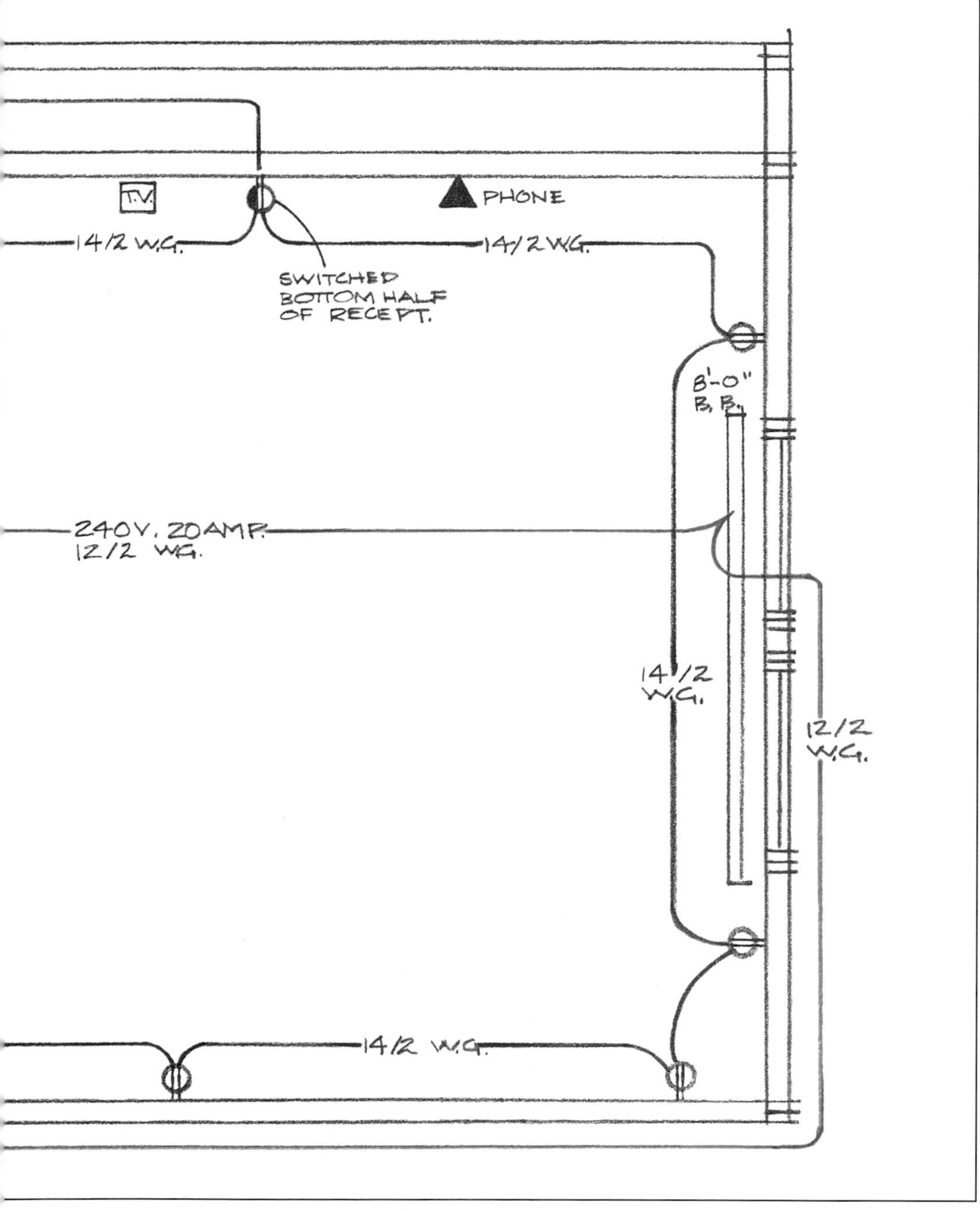

▶ **Telephone outlet:** 22-gauge four-wire phone cable (or eight-wire cable, if required by your telephone company), flush-mount telephone outlet.

Cable television jack: coaxial cable with F-connectors, signal splitter, cable television outlet with mounting brackets.

■ **Circuit #5 :** A 20-amp, 240-volt circuit that supplies power to three baseboard heaters controlled by a wall thermostat, and to a bathroom blower-heater controlled by a built-in thermostat. 12/2 NM cable, 750-watt blower-heater, single-gang box, line-voltage thermostat, three baseboard heaters, 20-amp double-pole circuit breaker.

Circuit #3: A 20-amp, 240-volt air-conditioner circuit. 12/2 NM cable; single-gang box; 20-amp, 240-volt receptacle (duplex or singleplex style); 20-amp double-pole circuit.

■ **Circuit #4 :** A 15-amp, 120-volt basic lighting/receptacle circuit serving most of the fixtures in the bedroom and study areas. 14/2 and 14/3 NM cable, 2 double-gang boxes, fan speed-control switch, dimmer switch, single-pole switch, 2 three-way switches, 2 plastic light fixture boxes, light fixture for stairway, smoke detector, metal light fixture box with brace bar, ceiling fan with light fixture, 10 single-gang boxes, 4" × 4" box with single-gang adapter plate, 10 duplex receptacles (15-amp), 15-amp single-pole circuit breaker.

1: Plan the Circuits

Your plans for wiring a room addition should reflect how you will use the space. For example, an attic space used as a bedroom requires an air-conditioner circuit, while a basement area used as a sewing room needs extra lighting. See pages 6 to 11 for information on planning circuits, and call or visit your city building inspector's office to learn the local Code requirements. You will need to create a detailed wiring diagram and a list of materials before the inspector will grant a work permit for your job.

The National Electrical Code requires receptacles to be spaced no more than 12 ft. apart, but for convenience you can space them as close as 6 ft. apart. Also consider the placement of furniture in the finished room, and do not place receptacles or baseboard heaters where beds, desks, or couches will cover them.

Electric heating units are most effective if you position them on the outside walls, underneath the windows. Position the receptacles to the sides of the heating units, not above the heaters where high temperatures might damage electrical cords.

Room light fixtures should be centered in the room, while stairway lights must be positioned so each step is illuminated. All wall switches should be within easy reach of the room entrance. Include a smoke alarm if your room addition includes a sleeping area.

Installing a ceiling fan improves heating and cooling efficiency and is a good idea for any room addition. Position it in a central location, and make sure there is plenty of headroom beneath it. Also consider adding accessory wiring for telephone outlets, television jacks, or stereo speakers.

Tips for Planning Room Addition Circuits

A bathroom vent fan (pages 22 to 23) may be required by your local Building Code, especially if your bathroom does not have a window. Vent fans are rated according to room size. Find the bathroom size in square feet by multiplying the length of the room times its width, and buy a vent fan rated for this size.

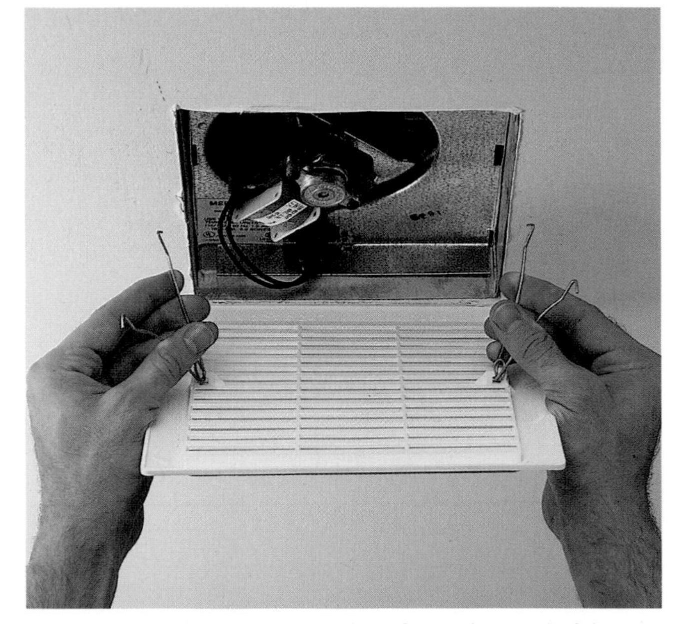

A permanently wired smoke alarm (page 31) is required by local Building Codes for room additions that include sleeping areas. Plan to install the smoke alarm just outside the sleeping area, in a hallway or stairway. Battery-operated smoke detectors are not allowed in new room additions.

If your room addition includes a bathroom, it will have special wiring needs. All bathrooms require one or more GFCI receptacles, and most need a vent fan. An electric blower-heater will make your bathroom more comfortable.

Before drawing diagrams and applying for a work permit, calculate the electrical load. Make sure your main service provides enough power for the new circuits.

Draw your wiring diagram. Using the completed diagram as a guide, create a detailed list of the materials you need. Bring the wiring diagram and the materials list to the inspector's office when you apply for the work permit. If the inspector suggests changes or improvements to your circuit design, follow his advice. His suggestions can save you time and money, and will ensure a safe wiring installation.

A wiring plan for a room addition should show the location of all partition walls, doorways, and windows. Mark the location of all new and existing plumbing fixtures, water lines, drains, and vent pipes. Draw in any chimneys and duct work for central heating and air-conditioning systems. Make sure the plan is drawn to scale, because the size of the space will determine how you route the electrical cables and arrange the receptacles and fixtures.

Blower-heaters with built-in thermostats (pages 20, 32) work well in small areas like bathrooms, where quick heat is important. Some models can be wired for either 120 or 240 volts. A bathroom blower-heater should be placed well away from the sink and tub, at a comfortable height where the controls are easy to reach. In larger rooms, electric baseboard heaters controlled by a wall thermostat are more effective than blower-heaters.

Telephone and cable television wiring (pages 24 to 25) is easy to install at the same time you are installing electrical circuits. Position the accessory outlets in convenient locations, and keep the wiring at least 6" away from the electrical circuits to prevent static interference.

2: Install Boxes & Cables

For efficiency, install the electrical boxes for all new circuits before running any of the cables. After all the cables are installed, your project is ready for the rough-in inspection. Do not make the final connections until your work has passed rough-in inspection.

Boxes: Your room addition may have recessed fixtures, like a blower-heater (photo, right) or vent fan (pages 22 to 23). These recessed fixtures have built-in wire connection boxes, and should be installed at the same time you are installing the standard electrical boxes. For ceiling fan or other heavy ceiling fixture, install a metal box and brace bar (page opposite).

Cables: You can install the necessary wiring for telephone outlets and cable television jacks (pages 24 to 25). The wiring is easy to install at the same time you are running electrical circuits, and is not subject to formal inspection.

How to Install a Blower-Heater

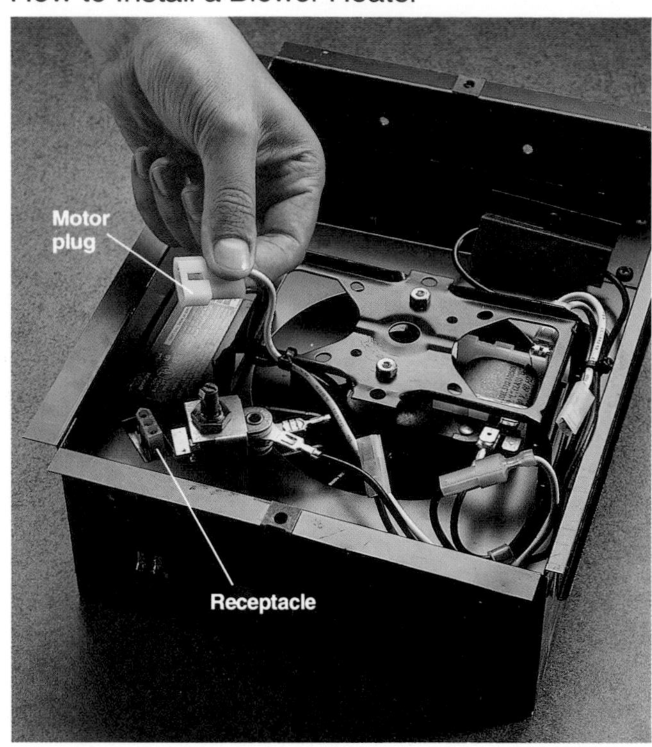

1 Disconnect the motor plug from the built-in receptacle that extends through the motor plate from the wire connection box.

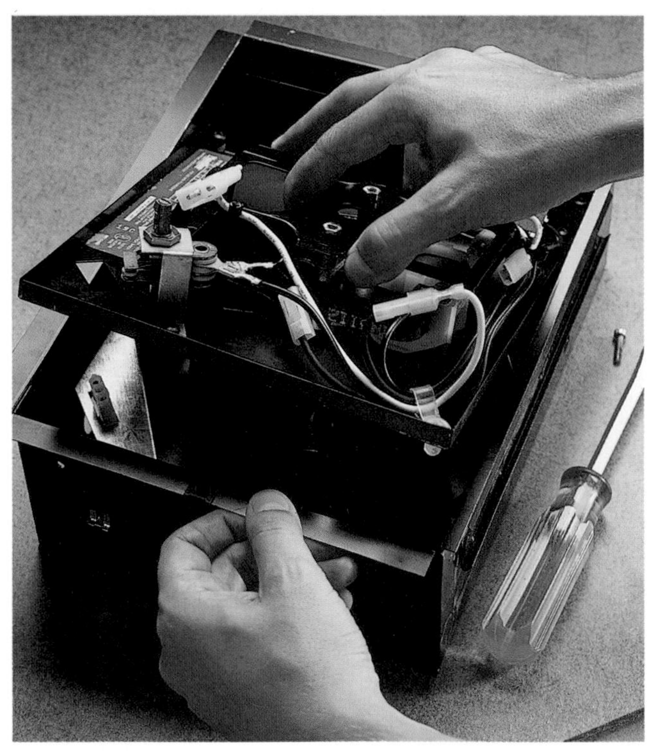

2 Take out the motor unit by removing the mounting screw and sliding the unit out of the frame.

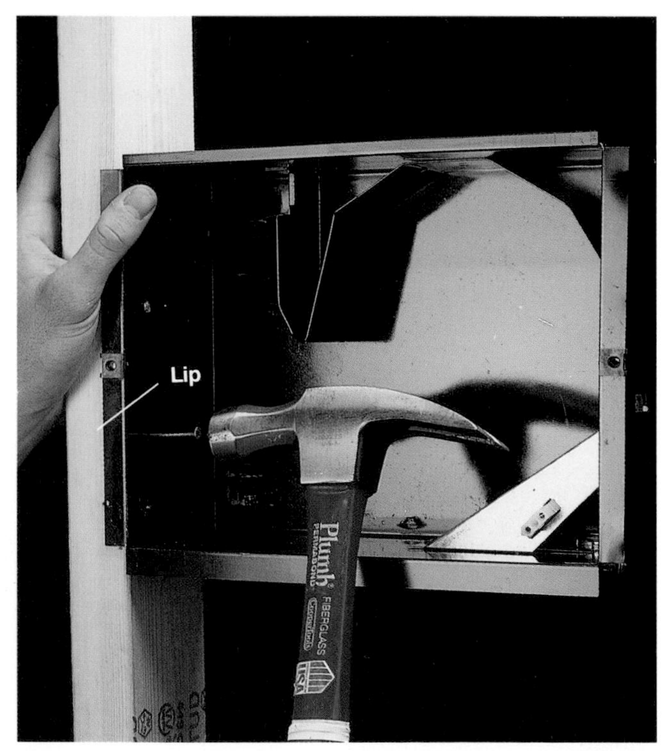

3 Open one knockout for each cable that will enter the wire connection box. Attach a cable clamp to each knockout. Position frame against a wall stud so the front lip will be flush with the finished wall surface. Attach the frame as directed by the manufacturer.

How to Install a Metal Box & Brace Bar for a Ceiling Fan

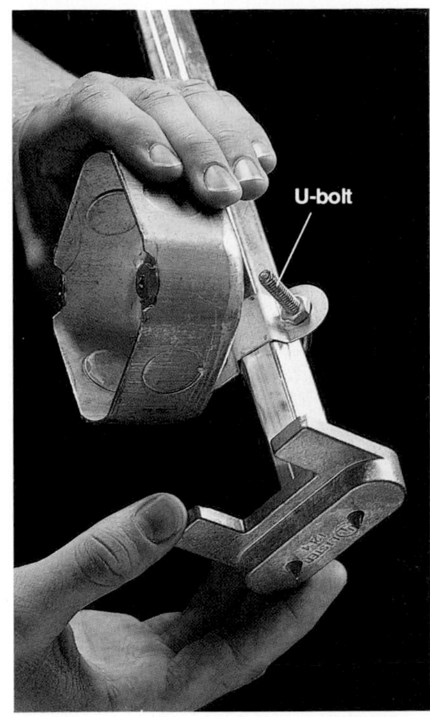

1 Attach a $1^1/_2$"-deep metal light fixture box to the brace bar, using a U-bolt and two nuts.

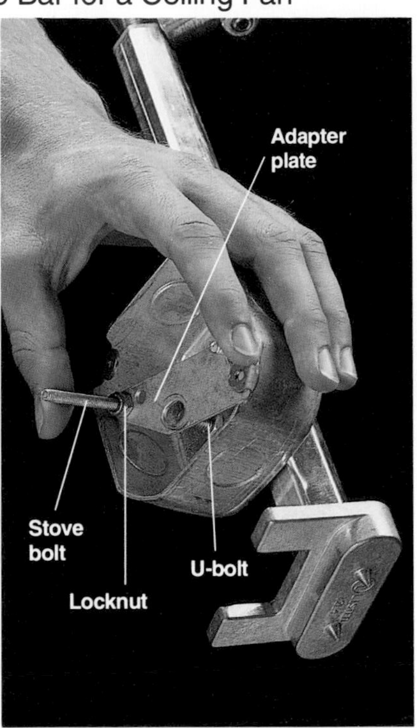

2 Attach the included stove bolts to the adapter plate with locknuts. These bolts will support the fan. Insert the adapter plate into the box so ends of U-bolt fit through the holes on the adapter plate.

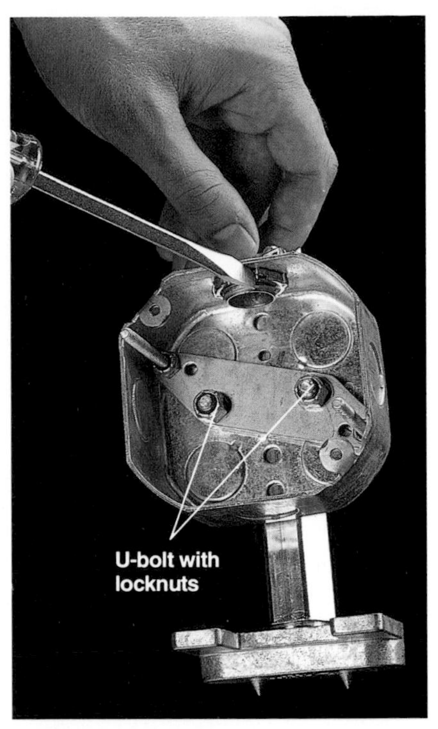

3 Secure the adapter plate by screwing two locknuts onto the U-bolt. Open one knockout for each cable that will enter the electrical box, and attach a cable clamp to each knockout.

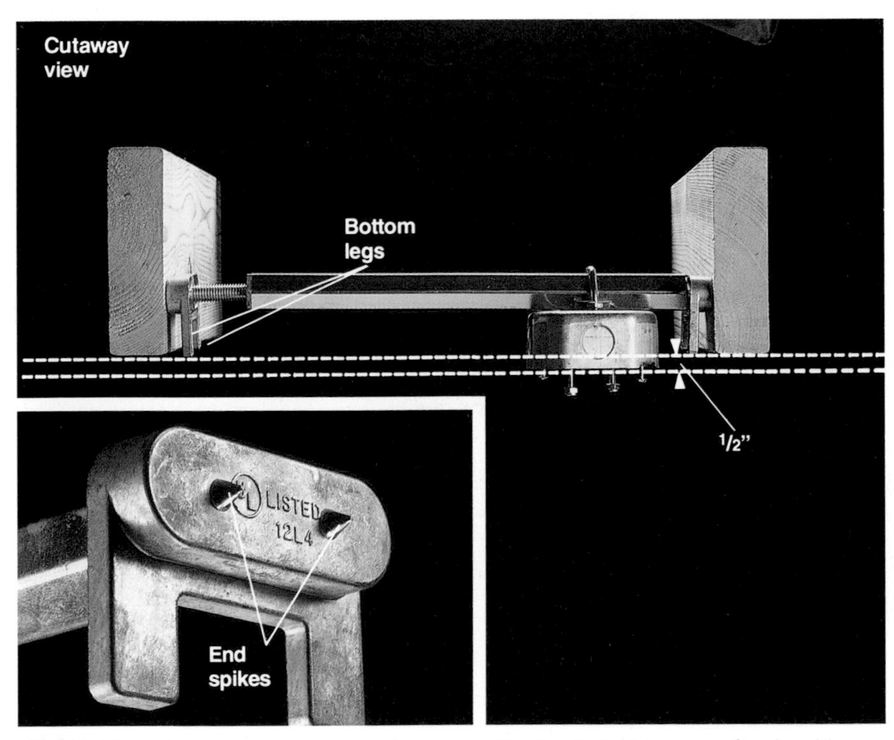

4 Position the brace between joists so the bottom legs are flush with the bottom of the joists. Rotate the bar by hand to force the end spikes into the joists. The face of the electrical box should be below the joists so the box will be flush with the finished ceiling surface.

5 Tighten the brace bar one rotation with a wrench to anchor the brace tightly against the joists.

Installing a Vent Fan

A vent fan helps prevent moisture damage to a bathroom by exhausting humid air to the outdoors. Vent fans are rated to match different room sizes. A vent fan can be controlled by a wall-mounted timer or single-pole switch. Some models have built-in light fixtures.

Position the vent fan in the center of the bathroom or over the stool area. In colder regions, Building Codes require that the vent hose be wrapped with insulation to prevent condensation of the moist air passing through the hose.

A vent fan has a built-in motor and blower that exhausts moisture-laden air from a bathroom to the outdoors through a plastic vent hose. A two-wire cable from a wall-mounted timer or single-pole switch is attached to the fan wire connection box with a cable clamp. A louvered coverplate mounted on the outside wall seals the vent against outdoor air when the motor is stopped.

How to Install a Vent Fan

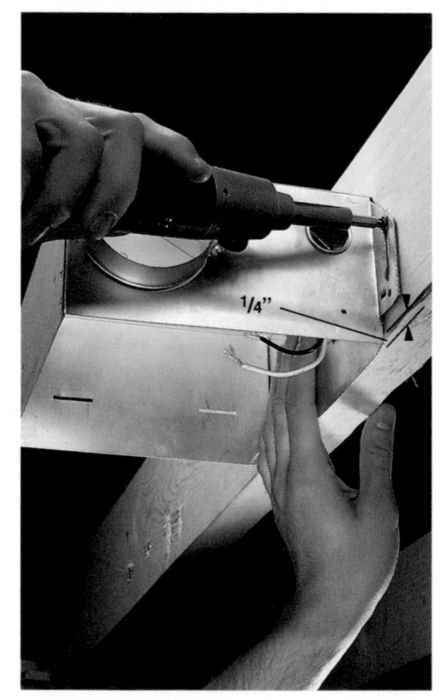

1 Disassemble the fan, following manufacturer's directions. Position the frame against a rafter so edge extends 1/4" below bottom edge of rafter to provide proper spacing for grill cover. Anchor frame with wallboard screws.

2 Choose the exit location for the vent. Temporarily remove any insulation, and draw the outline of the vent flange opening on the wall sheathing.

3 Drill a pilot hole, then make the cutout by sawing through the sheathing and siding with a jig saw. Keep the blade to the outside edge of the guideline.

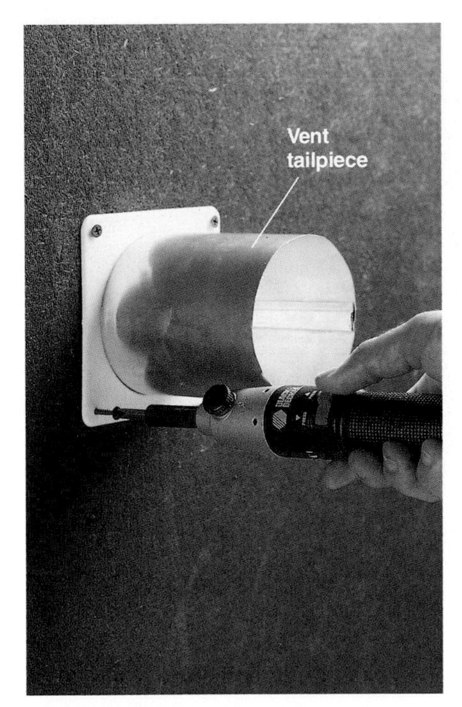

4 Insert the vent tailpiece into the cutout, and attach it to the wall by driving wallboard screws through the flange and into the sheathing.

5 Slide one end of vent hose over the tailpiece. Place one of the hose clamps around the end of the vent hose and tighten with a screwdriver. Replace insulation against sheathing.

6 Attach a hose adapter to the outlet on the fan frame by driving sheet-metal screws through the adapter and into the outlet flange. (NOTE: on some fans a hose adapter is not required.)

7 Slide the vent hose over the adapter. Place a hose clamp around the end of the hose and tighten it with a screwdriver. Your Building Code may require that you insulate the vent hose to prevent condensation problems.

8 On the outside wall of the house, place the louvered vent cover over the vent tailpiece, making sure the louvers are facing down. Attach the cover to the wall with galvanized screws. Apply a thick bead of caulk around the edge of the cover.

Arrange for the rough-in inspection before making the final connections.

Installing Telephone & Cable Television Wiring

Telephone outlets and television jacks are easy to install while you are wiring new electrical circuits. Install the accessory cables while framing members are exposed, then make the final connections after the walls are finished.

Telephone lines use four- or eight-wire cable, often called "bell wire," while television lines use a shielded coaxial cable with threaded end fittings called F-connectors. To splice into an existing cable television line, use a fitting called a signal splitter. Signal splitters are available with two, three, or four outlet nipples.

How to Install Coaxial Cable for a Television Jack

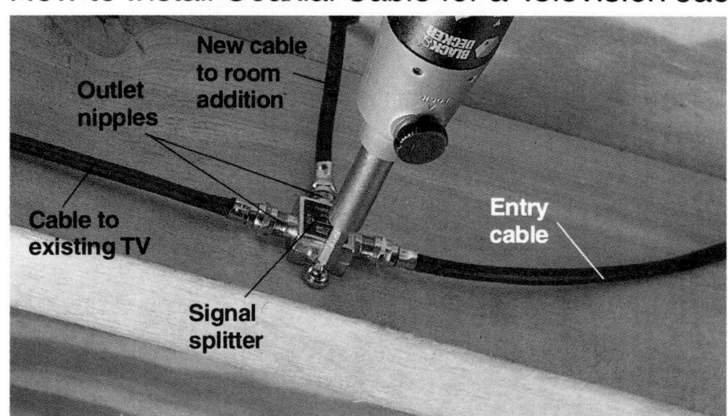

1 Install a signal splitter where the entry cable connects to indoor TV cables, usually in the basement or another utility area. Attach one end of new coaxial cable to an outlet nipple on the splitter. Anchor splitter to a framing member with wallboard screws.

2 Run the coaxial cable to the location of the new television jack. Keep coaxial cable at least 6" away from electrical wiring to avoid electrical interference. Mark the floor so the cable can be found easily after the walls are finished.

How to Connect a Television Jack

1 After walls are finished, make a cutout opening $1^1/_2$" wide and $3\ ^3/_4$" high at the television jack location. Pull cable through the opening, and install two television jack mounting brackets in the cutout.

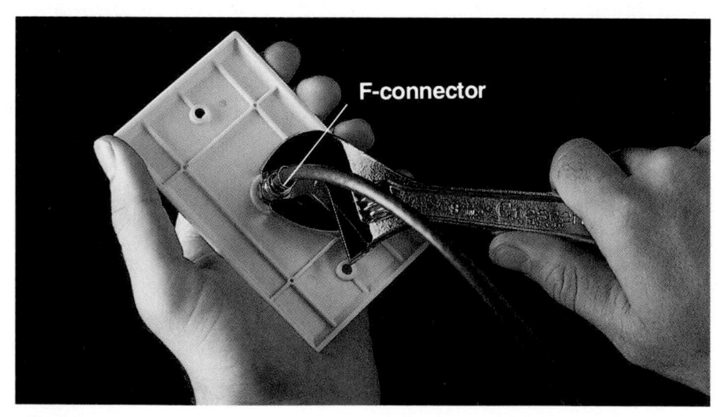

2 Use a wrench to attach the cable F-connector to the back of the television jack. Attach the jack to the wall by screwing it onto the mounting brackets.

How to Install Cable for a Telephone Outlet

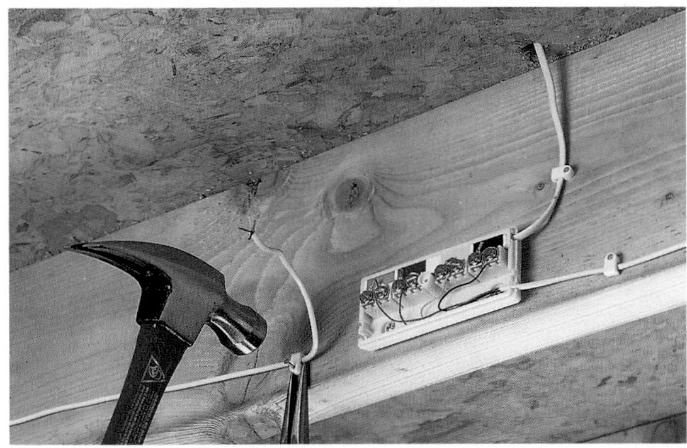

1 Locate a telephone junction in your basement or other utility area. Remove the junction cover. Use cable staples to anchor one end of the cable to a framing member near the junction, leaving 6" to 8" of excess cable.

2 Run the cable from the junction to the telephone outlet location. Keep the cable at least 6" away from circuit wiring to avoid electrical interference. Mark the floor so the cable can be located easily after the walls are finished.

How to Connect a Telephone Outlet

1 After walls are finished, cut a hole in the wallboard at phone outlet location, using a wallboard saw. Retrieve the cable, using a piece of stiff wire.

2 At each cable end, remove about 2" of outer sheathing. Remove about 3/4" of insulation from each wire, using a combination tool.

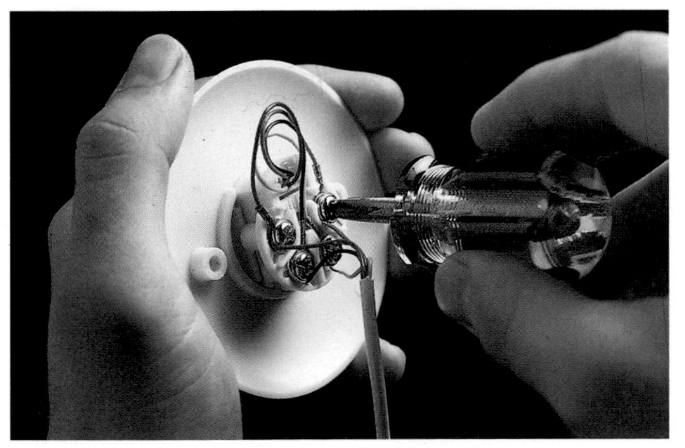

3 Connect wires to similarly colored wire leads in phone outlet. If there are extra wires, tape them to back of outlet. Put the telephone outlet over the wall cutout, and attach it to the wallboard.

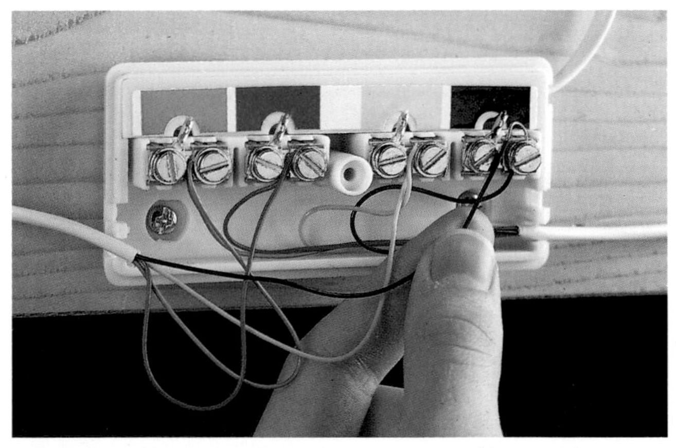

4 At the telephone junction, connect the cable wires to the color-coded screw terminals. If there are extra wires, wrap them with tape and tuck them inside the junction. Reattach the junction cover.

3: Make Final Connections

Make the final connections for receptacles, switches, and fixtures only after the rough-in inspection is done and all walls and ceilings are finished. The last step is to hook up the new circuits at the breaker panel.

After all connections are done, your work is ready for the final inspection. If you have worked carefully, the final inspection will take only a few minutes. The inspector may open one or two electrical boxes to check wire connections, and will check the circuit breaker hookups to make sure they are correct.

Materials You Will Need:

Pigtail wires, wire nuts, green & black tape.

Circuit #1

A 15-amp, 120-volt circuit serving the bathroom & closet.

- Timer & single-pole switch
- Vent fan
- Two light fixtures
- GFCI receptacle
- Single-pole switch
- 15-amp single-pole circuit breaker

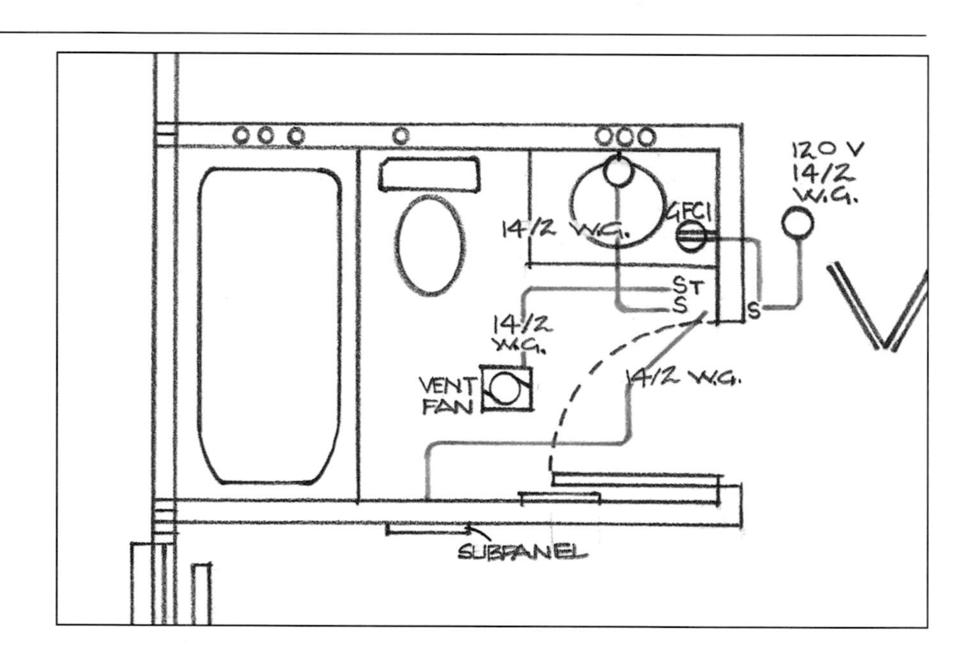

How to Connect the Timer & Single-pole Switch

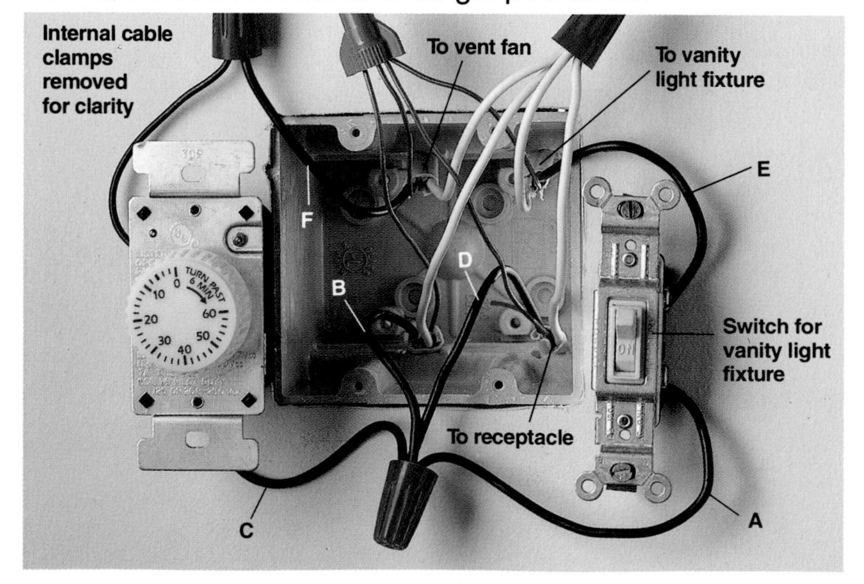

Attach a black pigtail wire (A) to one of the screw terminals on the switch. Use a wire nut to connect this pigtail to the black feed wire (B), to one of the black wire leads on the timer (C), and to the black wire carrying power to the bathroom receptacle (D). Connect the black wire leading to the vanity light fixture (E) to the remaining screw terminal on the switch. Connect the black wire running to the vent fan (F) to the remaining wire lead on the timer. Use wire nuts to join the white wires and the grounding wires. Tuck all wires into the box, then attach the switches, coverplate and timer dial.

How to Connect the Vent Fan

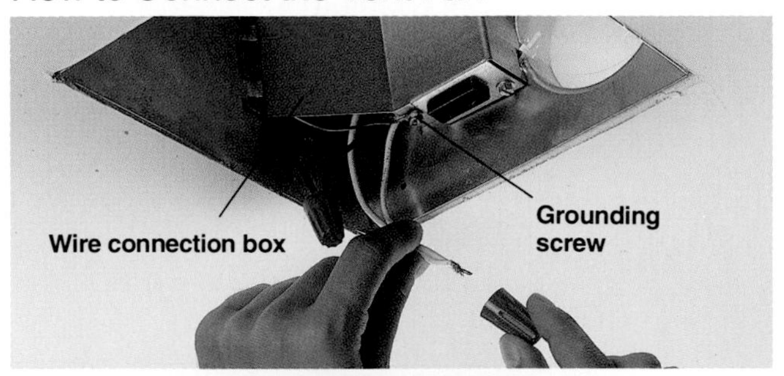

In the wire connecting box (top) connect black circuit wire to black wire lead on fan, using a wire nut. Connect white circuit wire to white wire lead. Connect grounding wire to the green grounding screw. **Insert the fan motor unit** (bottom) and attach mounting screws. Connect the fan motor plug to the built-in receptacle on the wire connection box. Attach the fan grill to the frame, using the mounting clips included with the fan kit (page 22).

How to Connect Light Fixtures

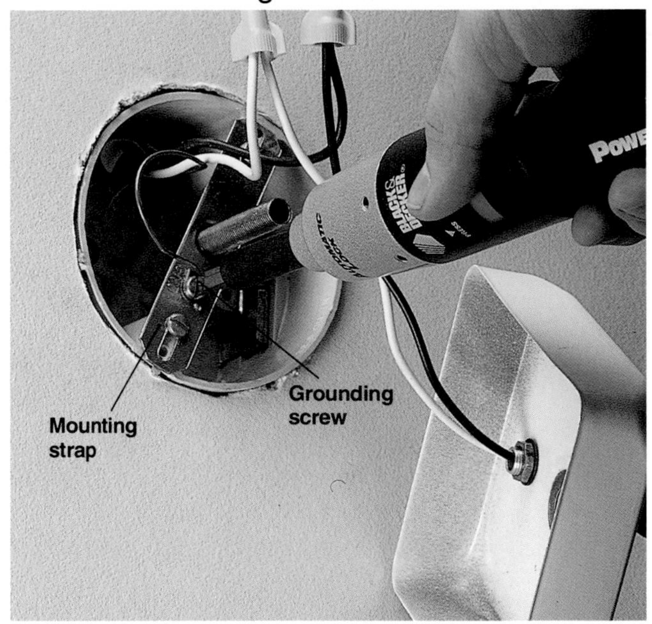

Attach a mounting strap with threaded nipple to the box, if required by the light fixture manufacturer. Connect the black circuit wire to the black wire lead on the light fixture, and connect the white circuit wire to the white wire lead. Connect the circuit grounding wire to the grounding screw on the mounting strap. Carefully tuck all wires into the electrical box, then position the fixture over the nipple and attach it with the mounting nut.

How to Connect the Bathroom GFCI Receptacle

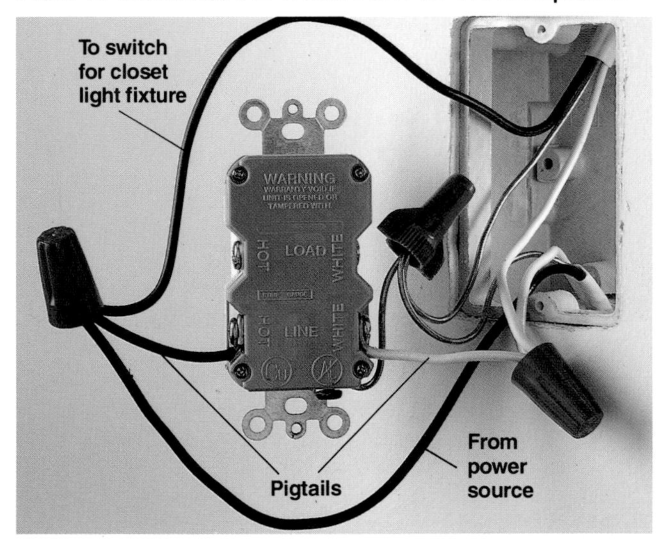

Attach a black pigtail wire to brass screw terminal marked LINE. Join all black wires with a wire nut. Attach a white pigtail wire to the silver screw terminal marked LINE, then join all white wires with a wire nut. Attach a grounding pigtail to the green grounding screw, then join all grounding wires. Tuck all wires into the box, then attach the receptacle and the coverplate.

How to Connect the Single-pole Switch

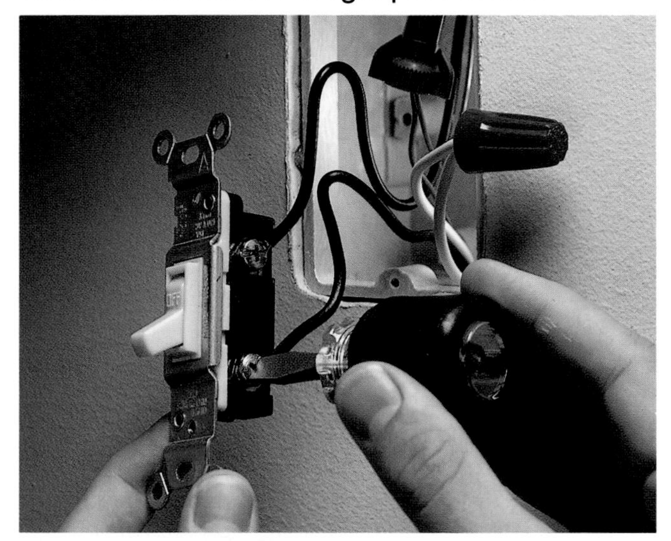

Attach the black circuit wires to brass screw terminals on the switch. Use wire nuts to join the white neutral wires together and the bare copper grounding wires together. Tuck all wires into the box, then attach the switch and the coverplate.

Circuit #2:

A 15-amp, 120-volt isolated-ground circuit for a home computer in the office area.

- 15-amp isolated-ground receptacle
- 15-amp single-pole circuit breaker

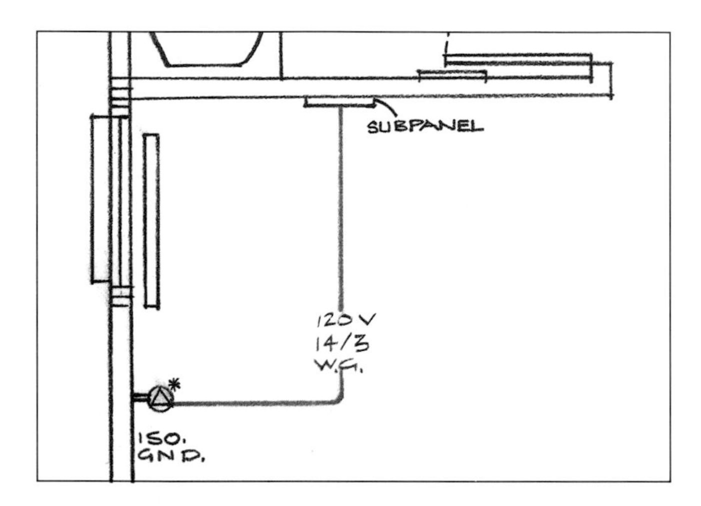

How to Connect the Computer Receptacle

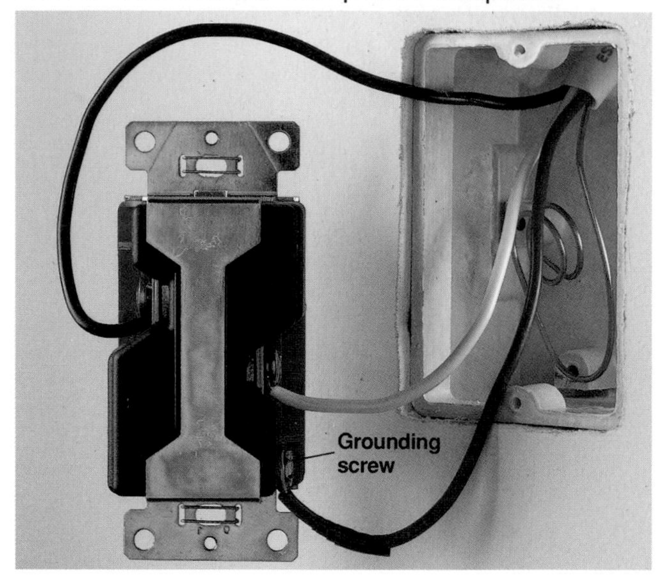

Tag the red wire with green tape to identify it as a grounding wire. Attach this wire to the grounding screw terminal on the isolated-ground receptacle. Connect the black wire to the brass screw terminal, and the white wire to the silver screw. Push the bare copper wire to the back of the box. Carefully tuck all wires into the box, then attach the receptacle and coverplate.

Circuit #3:

A 20-amp, 240-volt air-conditioner circuit.

- 20-amp, 240-volt receptacle (singleplex or duplex style)
- 20-amp double-pole circuit breaker

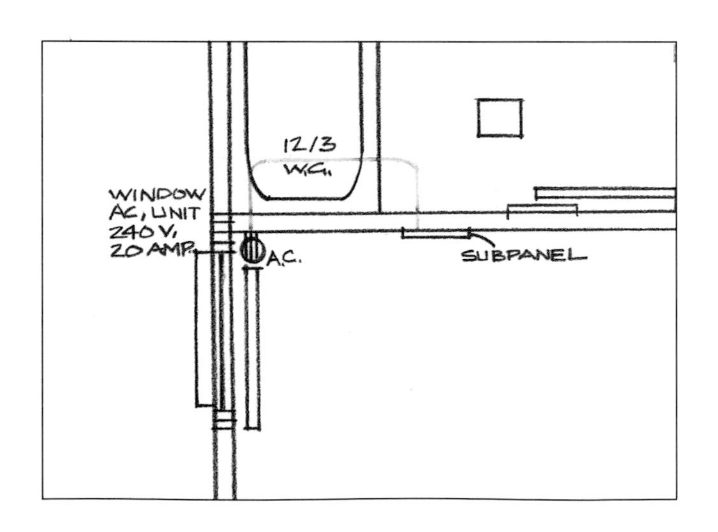

How to Connect the 240-volt Receptacle

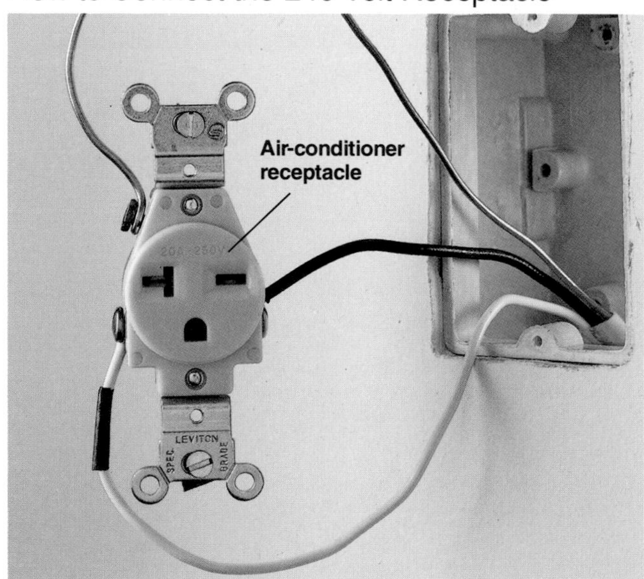

Connect the black circuit wire to a brass screw terminal on the air-conditioner receptacle, and connect the white circuit wire to the screw on the opposite side. Tag white wire with black tape to identify it as a hot wire. Connect grounding wire to green grounding screw on the receptacle. Tuck in wires, then attach the receptacle and coverplate. A 240-volt receptacle is available in either singleplex (shown above) or duplex style.

Circuit #4:

A 15-amp, 120-volt basic lighting/receptacle circuit serving the office and bedroom areas.

- Single-pole switch for split receptacle, three-way switch for stairway light fixture
- Speed-control and dimmer switches for ceiling fan
- Switched duplex receptacle
- 15-amp, 120-volt receptacles
- Ceiling fan with light fixture
- Smoke detector
- Stairway light fixture
- 15-amp single-pole circuit breaker

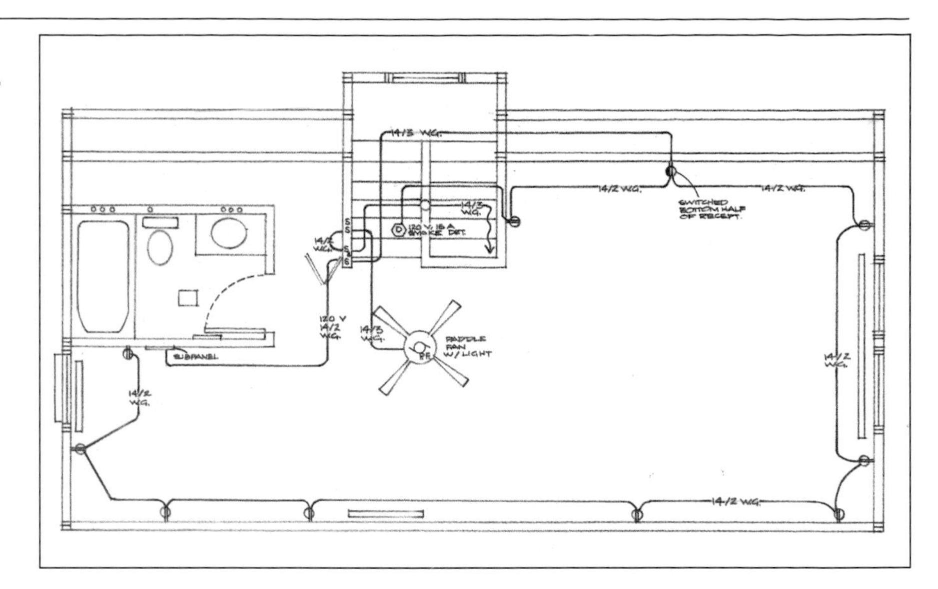

How to Connect Switches for Receptacle & Stairway Light

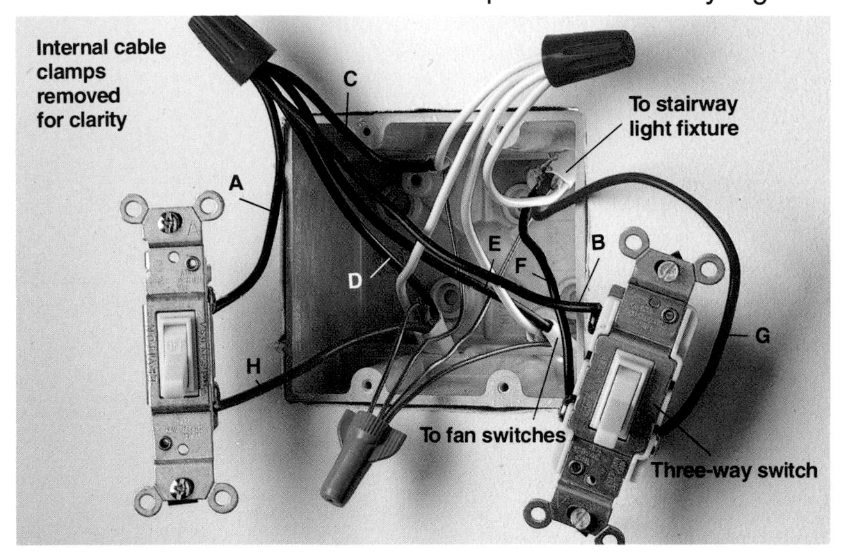

Attach a black pigtail wire (A) to one of the screws on the single-pole switch and another black pigtail (B) to common screw on three-way switch. Use a wire nut to connect pigtail wires to black feed wire (C), to black wire running to unswitched receptacles (D), and to the black wire running to fan switches (E). Connect remaining wires running to light fixture (F, G) to traveler screws on three-way switch. Connect red wire running to switched receptacle (H) to remaining screw on single-pole switch. Use wire nuts to join white wires and grounding wires. Tuck all wires into box, then attach switches and coverplate.

How to Connect the Ceiling Fan Switches

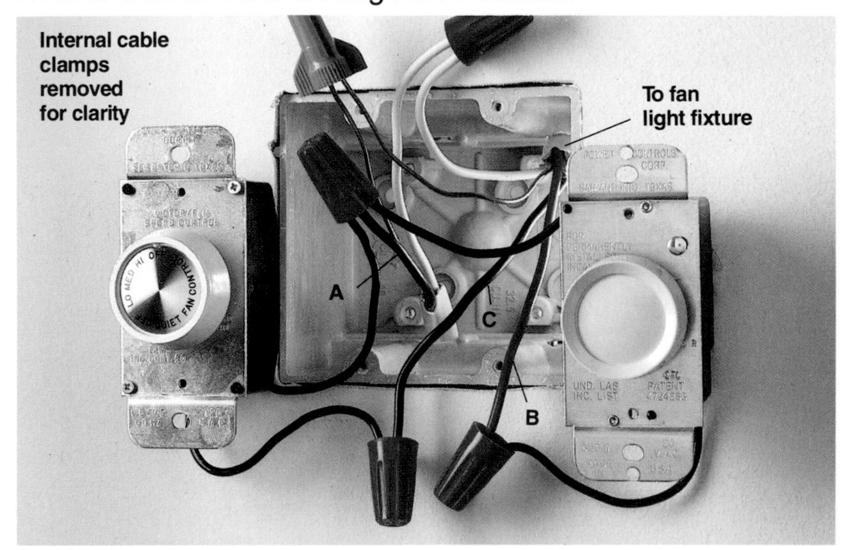

Connect the black feed wire (A) to one of the black wire leads on each switch, using a wire nut. Connect the red circuit wire (B) running to the fan light fixture to the remaining wire lead on the dimmer switch. Connect the black circuit wire (C) running to the fan motor to the remaining wire lead on the speed-control switch. Use wire nuts to join the white wires and the grounding wires. Tuck all wires into the box, then attach the switches, coverplate, and switch dials.

How to Connect a Switched Receptacle

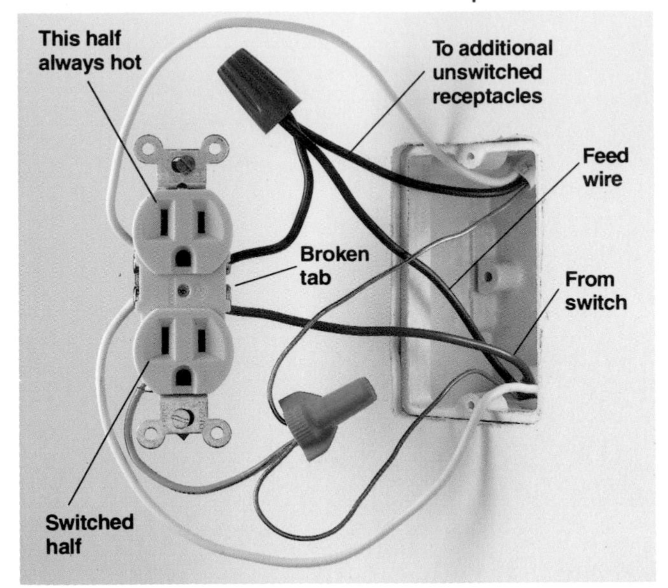

Break the connecting tab between the brass screw terminals on the receptacle, using needlenose pliers. Attach the red wire to the bottom brass screw. Connect a black pigtail wire to the other brass screw, then connect all black wires with wire nut. Connect white wires to silver screws. Attach a grounding pigtail to the green grounding screw, then join all the grounding wires, using a wire nut. Tuck the wires into the box, then attach the receptacle and coverplate.

How to Connect Receptacles

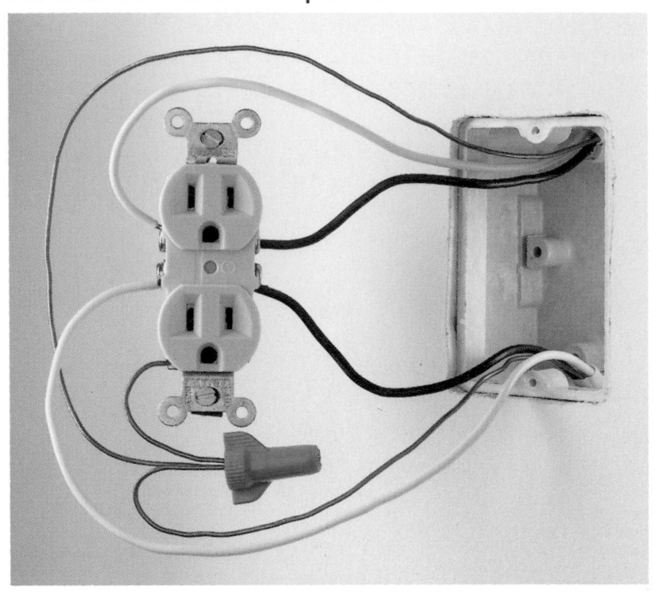

Connect the black circuit wires to the brass screw terminals on the receptacle, and the white wires to the silver terminals. Attach a grounding pigtail to the green grounding screw on the receptacle, then join all grounding wires with a wire nut. Tuck the wires into the box, then attach the receptacle and the coverplate.

How to Connect a Ceiling Fan/Light Fixture

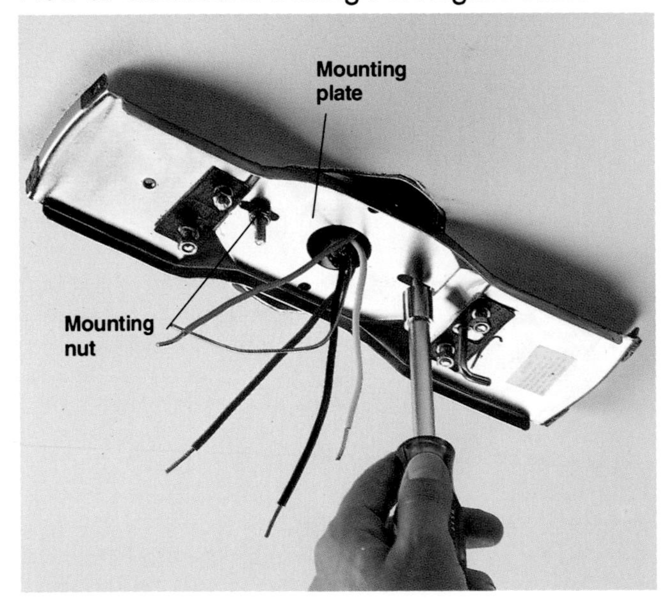

1 Place the ceiling fan mounting plate over the stove bolts extending through the electrical box. Pull the circuit wires through the hole in the center of the mounting plate. Attach the mounting nuts and tighten them with a nut driver.

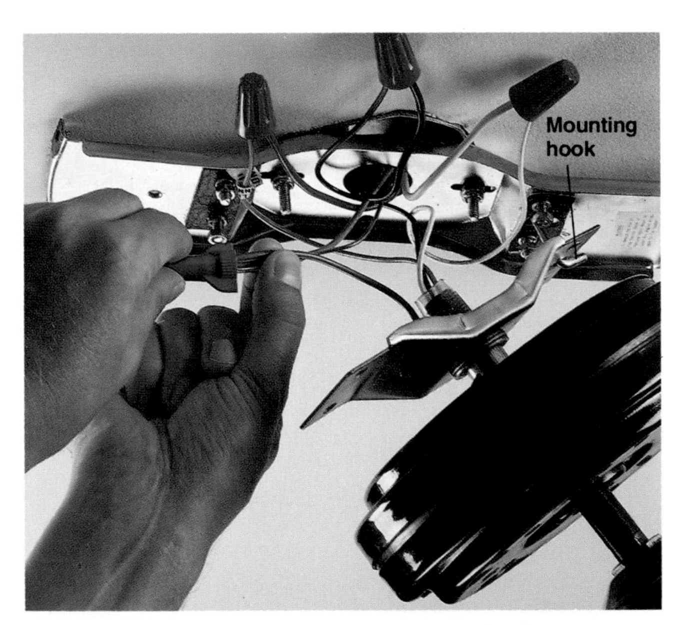

2 Hang fan motor from mounting hook. Connect black circuit wire to black wire lead from fan, using a wire nut. Connect red circuit wire from dimmer to blue wire lead from light fixture, white circuit wire to white lead, and grounding wires to green lead. Complete assembly of fan and light fixture, following manufacturer's directions.

How to Connect a Smoke Alarm

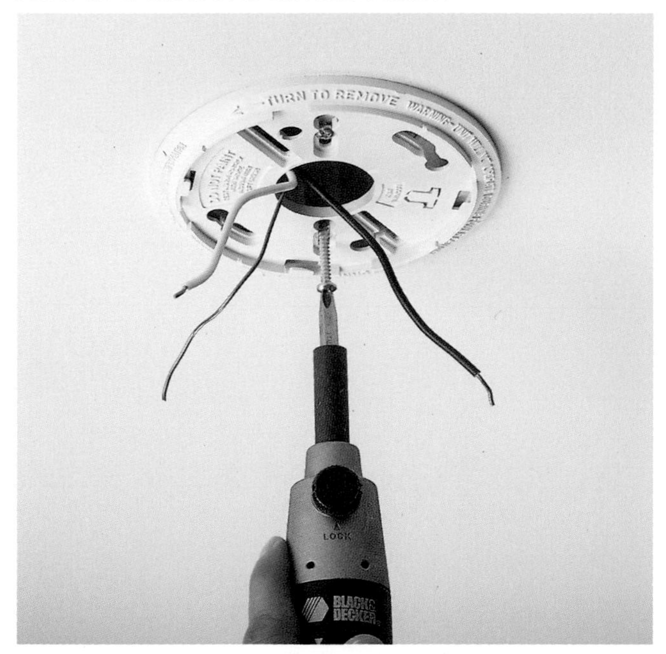

1 Attach the smoke alarm mounting plate to the electrical box, using the mounting screws provided with the smoke alarm kit.

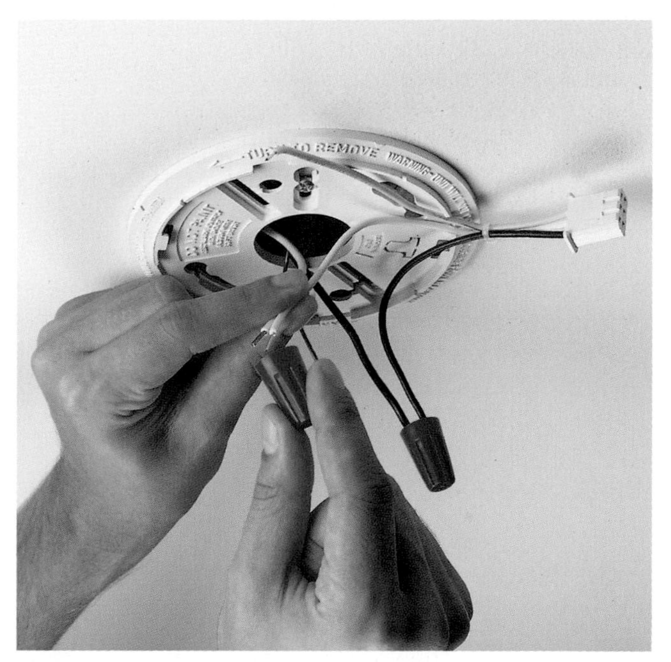

2 Use wire nuts to connect the black circuit wire to the black wire lead on the smoke alarm, and the white circuit wire to the white wire lead.

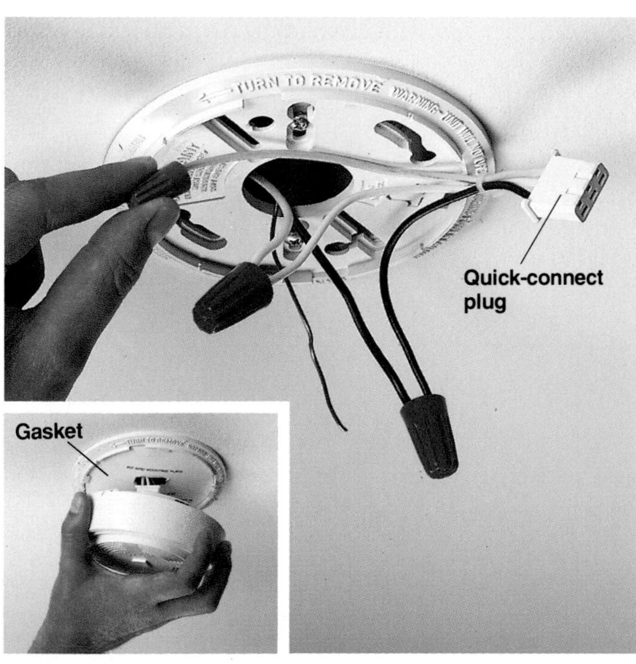

3 Screw a wire nut onto the end of the yellow wire, if present. (This wire is used only if two or more alarms are wired in series.) Tuck all wires into the box Place the cardboard gasket over the mounting plate. Attach the quick-connect plug to the smoke alarm. Attach the alarm to the mounting plate, twisting it clockwise until it locks into place (inset).

How to Connect a Stairway Light Fixture

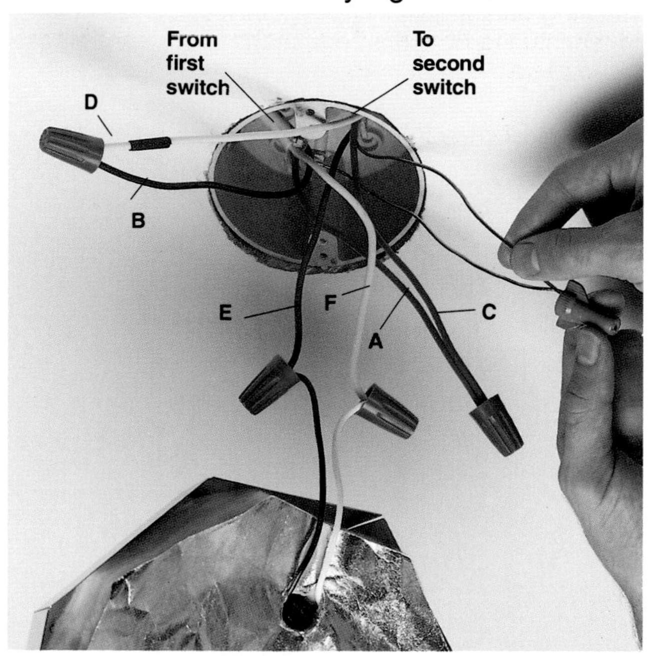

Connect the traveler wires entering the box from the first three-way switch (red wire [A] and black wire [B]) to the traveler wires running to the second three-way switch (red wire [C] and white wire tagged with black tape [D]). Connect the common wire running to the second switch (E) to the black lead on the light fixture. Connect the white wire from the first switch (F) to the white fixture lead. Join the grounding wires. Tuck wires into box and attach the light fixture.

Circuit #5:

A 20-amp, 240-volt circuit serving the bathroom blower-heater, and three baseboard heaters controlled by a wall thermostat.

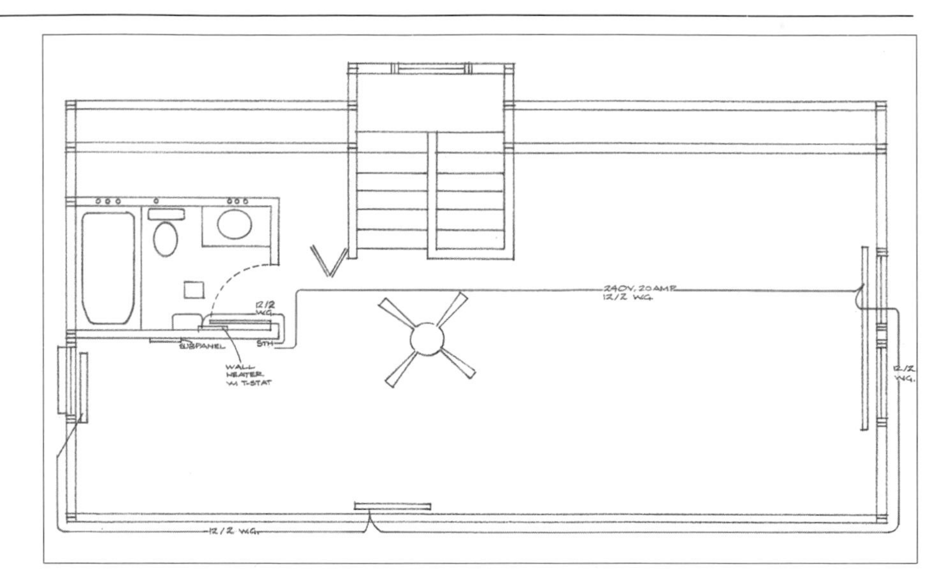

- 240-volt blower-heater
- 240-volt thermostat
- 240-volt baseboard heaters
- 20-amp double-pole circuit breaker

How to Connect a 240-volt Blower-Heater

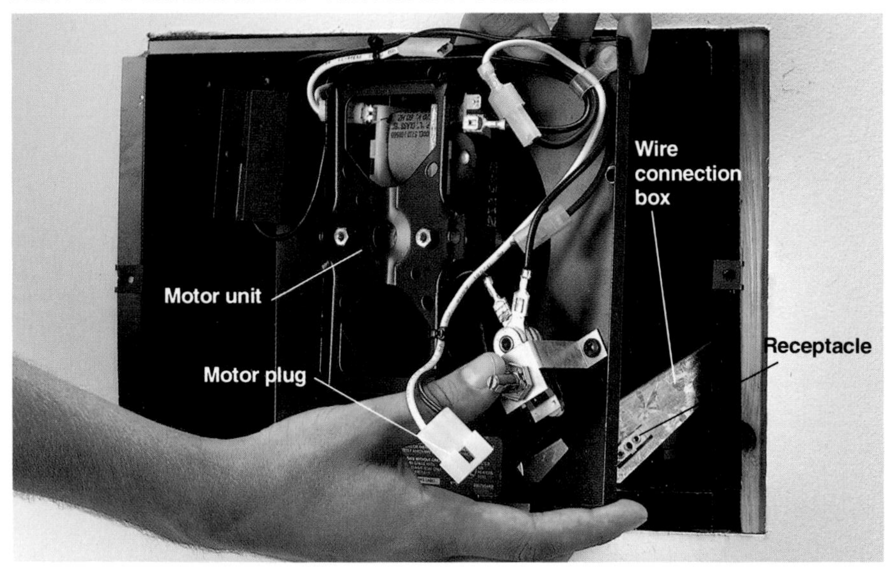

Blower-heaters: In the heater's wire connection box, connect one of the wire leads to the white circuit wires, and the other wire lead to the black circuit wires, using same method as for baseboard heaters (page opposite). Insert the motor unit, and attach the motor plug to the built-in receptacle. Attach the coverplate and thermostat knob. NOTE: Some types of blower-heaters can be wired for either 120 volts or 240 volts. If you have this type, make sure the internal plug connections are configured for 240 volts.

How to Connect a 240-volt Thermostat

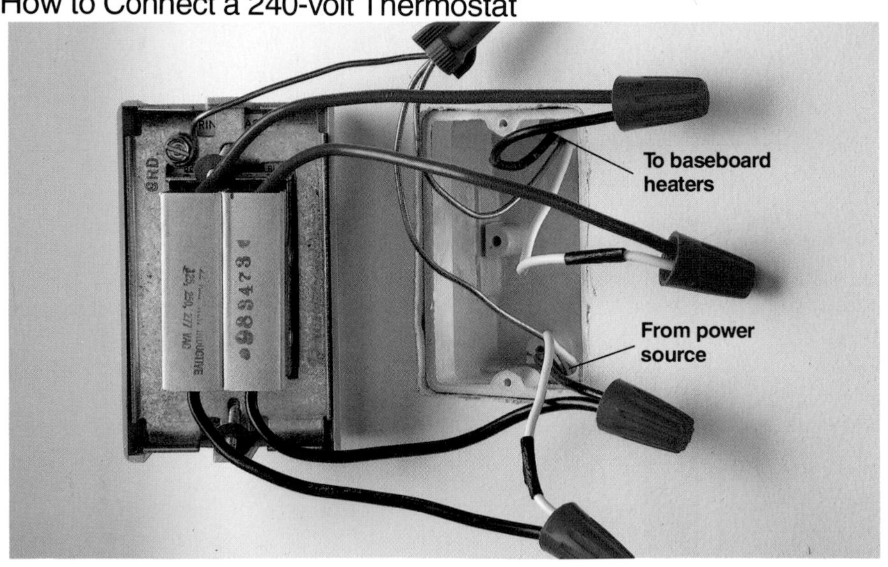

Connect the red wire leads on the thermostat to the circuit wires entering the box from the power source, using wire nuts. Connect black wire leads to circuit wires leading to the baseboard heaters. Tag the white wires with black tape to indicate they are hot. Attach a grounding pigtail to the grounding screw on the thermostat, then connect all grounding wires. Tuck the wires into the box, then attach the thermostat and coverplate. Follow manufacturer's directions: the color coding for thermostats may vary.

How to Connect 240-volt Baseboard Heaters

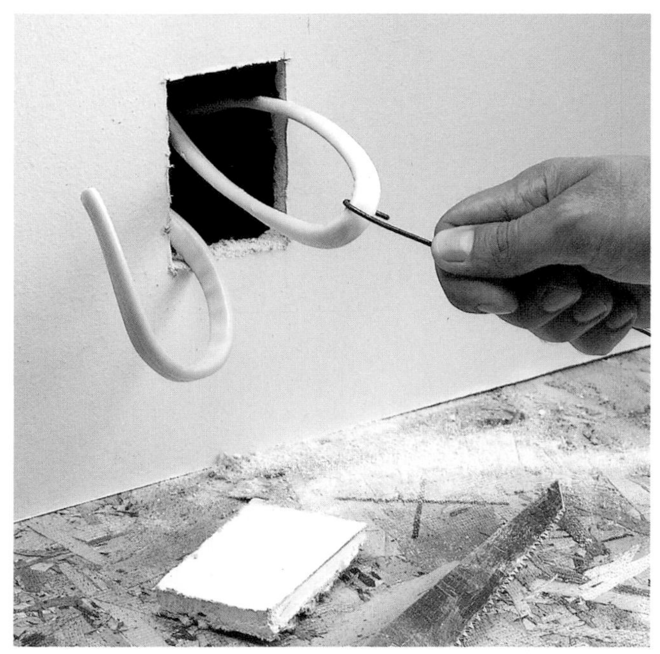

1 At the cable location, cut a small hole in the wallboard, 3" to 4" above the floor, using a wallboard saw. Pull the cables through the hole, using a piece of stiff wire with a hook on the end. Middle-of-run heaters will have 2 cables, while end-of-run heaters have only 1 cable.

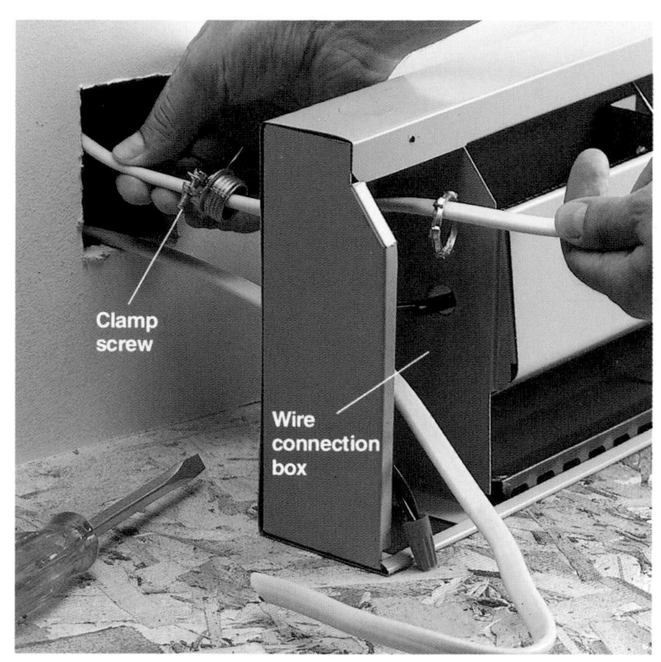

2 Remove the cover on the wire connection box. Open a knockout for each cable that will enter the box, then feed the cables through the cable clamps and into the wire connection box. Attach the clamps to the wire connection box, and tighten the clamp screws until the cables are gripped firmly.

3 Anchor heater against wall, about 1" off floor, by driving flat-head screws through back of housing and into studs. Strip away cable sheathing so at least 1/4" of sheathing extends into the heater. Strip 3/4" of insulation from each wire, using a combination tool.

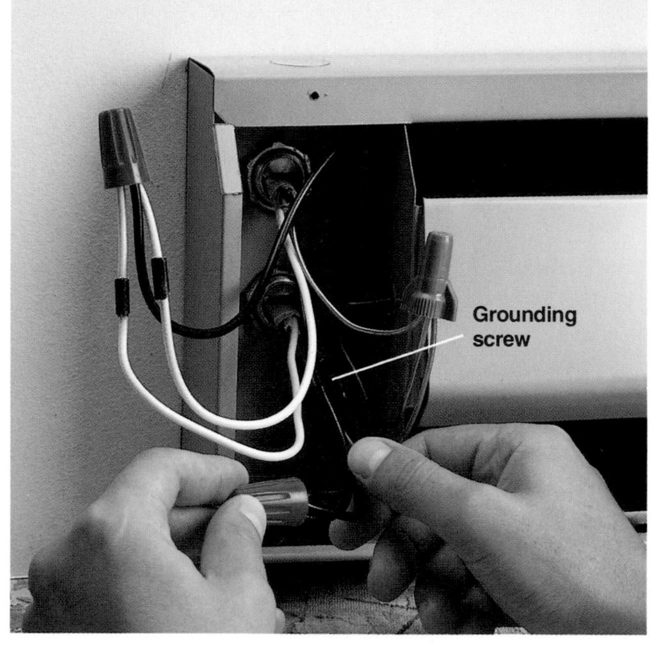

4 Use wire nuts to connect the white circuit wires to one of the wire leads on the heater. Tag white wires with black tape to indicate they are hot. Connect the black circuit wires to the other wire lead. Connect a grounding pigtail to the green grounding screw in the box, then join all grounding wires with a wire nut. Reattach cover.

Make hookups at circuit breaker panel and arrange for final inspection.

Choose the Fixtures You Need

A: Range receptacle (circuit #3) supplies power for a range/ oven combination appliance on a dedicated circuit. See page 48.

B: 20-amp receptacles (circuits #1 & #2) supply power for small appliances. See page 46.

C: Under-cabinet task lights (circuit #7) provide fluorescent light for countertop work areas. See page 51.

D: Microwave receptacle (circuit #4) supplies power for a microwave on a dedicated circuit. See page 48.

E: GFCI receptacles (circuits #1 & #2) provide protection against shock. See page 47.

Photo courtesy of Kitchens by Krengel, Inc.

Wiring a Remodeled Kitchen

The kitchen is the greatest power user in your home. Adding new circuits during a kitchen remodeling project will make your kitchen better serve your needs. This section shows how to install new circuit wiring when remodeling. You learn how to plan for the many power requirements of the modern kitchen, and techniques for doing the work before the walls and ceiling are finished.

This section takes you through all phases of the project: evaluating your existing service, planning the new work and getting a permit, installing the circuits, and having your work inspected.

You learn how to install circuits and fixtures for recessed lights, under-cabinet task lights, and a ceiling light controlled by three-way switches. You also learn how to install circuits and receptacles for a range, microwave, dishwasher, and food disposer. Methods for installing two small-appliance circuits are also shown.

While your kitchen remodeling project will differ from this one, the methods and concepts shown apply to any kitchen wiring project containing any combination of circuits.

The next two pages show the circuits in place with the walls and ceiling removed.

F: Ceiling fixture (circuit #7) provides general lighting for the entire kitchen. It is controlled by two three-way switches located by the doors to the room. See page 50.

G: Food disposer receptacle (circuit #5) is controlled by a switch near the sink and supplies power to the disposer located in the sink cabinet. See page 49.

H: Dishwasher receptacle (circuit #6) supplies power for the dishwasher on a dedicated circuit. See page 49.

I: Recessed fixtures (circuit #7) controlled by switches near the sink provide additional lighting for work areas at sink, range, and countertop. See page 51.

Learn How to Install These Circuits

#1 & #2: Small-appliance circuits. Two 20-amp, 120-volt circuits supply power to countertop and eating areas for small appliances. All general-use receptacles must be on these circuits. One 12/3 cable fed by a 20-amp double-pole breaker, wires both circuits. These circuits share one electrical box with the disposer circuit (#5), and another with the basic lighting circuit (#7).

#3: Range circuit. A 50-amp, 120/240-volt dedicated circuit supplies power to the range/oven appliance. It is wired with 6/3 cable.

#4: Microwave circuit. A dedicated 20-amp, 120-volt circuit supplies power to the microwave. It is wired with 12/2 cable. Microwaves that use less than 300 watts can be installed on a 15-amp circuit, or plugged into the small-appliance circuits.

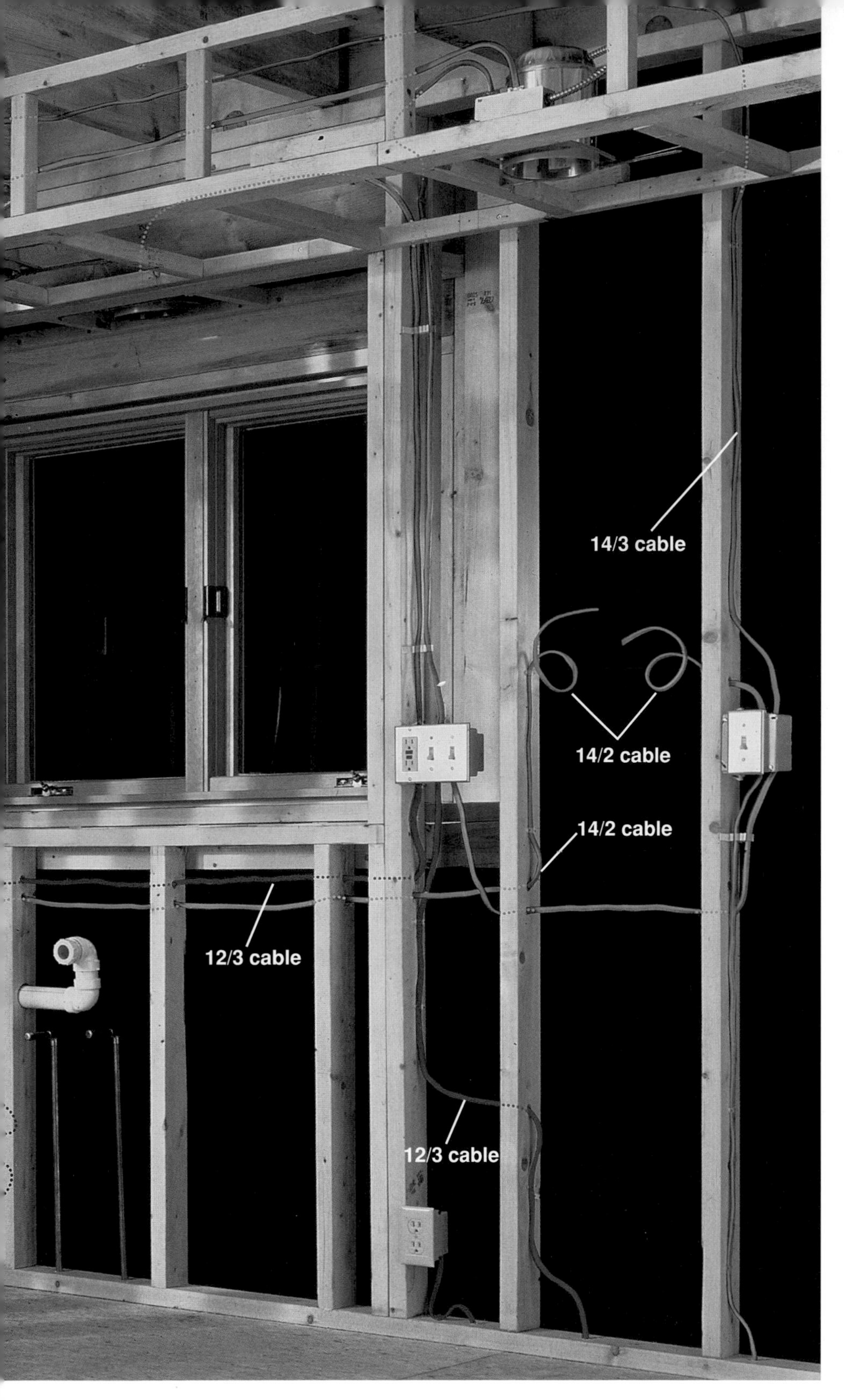

#5: Food disposer circuit. A dedicated 15-amp, 120-volt circuit supplies power to the disposer. It is wired with 14/2 cable. Some local Codes allow the disposer to be on the same circuit as the dishwasher.

#6: Dishwasher circuit. A dedicated 15-amp, 120-volt circuit supplies power to the dishwasher. It is wired with 14/2 cable. Some local Codes allow the dishwasher to be on the same circuit as the disposer.

#7: Basic lighting circuit. A 15-amp, 120-volt circuit powers the ceiling fixture, recessed fixtures, and under-cabinet task lights. 14/2 and 14/3 cables connect the fixtures and switches in the circuit. Each task light has a self-contained switch.

Wiring a Remodeled Kitchen: Construction View

The kitchen remodeling wiring project shown on the following pages includes the installation of seven new circuits. Four of these are dedicated circuits: a 50-amp circuit supplying the range, a 20-amp circuit powering the microwave, and two 15-amp circuits supplying the dishwasher and food disposer. In addition, two 20-amp circuits for small appliances supply power to all receptacles above the countertops and in the eating area. Finally, a 15-amp basic lighting circuit controls the ceiling fixture, all of the recessed fixtures, and the under-cabinet task lights.

All rough construction and plumbing work should be finished and inspected before beginning the electrical work. Divide the project into steps and complete each step before beginning the next.

Three Steps for Wiring a Remodeled Kitchen:

1. Plan the circuits (pages 40 to 41).
2. Install boxes and cables (pages 42 to 45).
3. Make final connections (pages 46 to 51).

Tools You Will Need:
Marker, tape measure, calculator, masking tape, screwdriver, hammer, power drill with 5/8" spade bit, cable ripper, combination tool, needlenose pliers, fish tape.

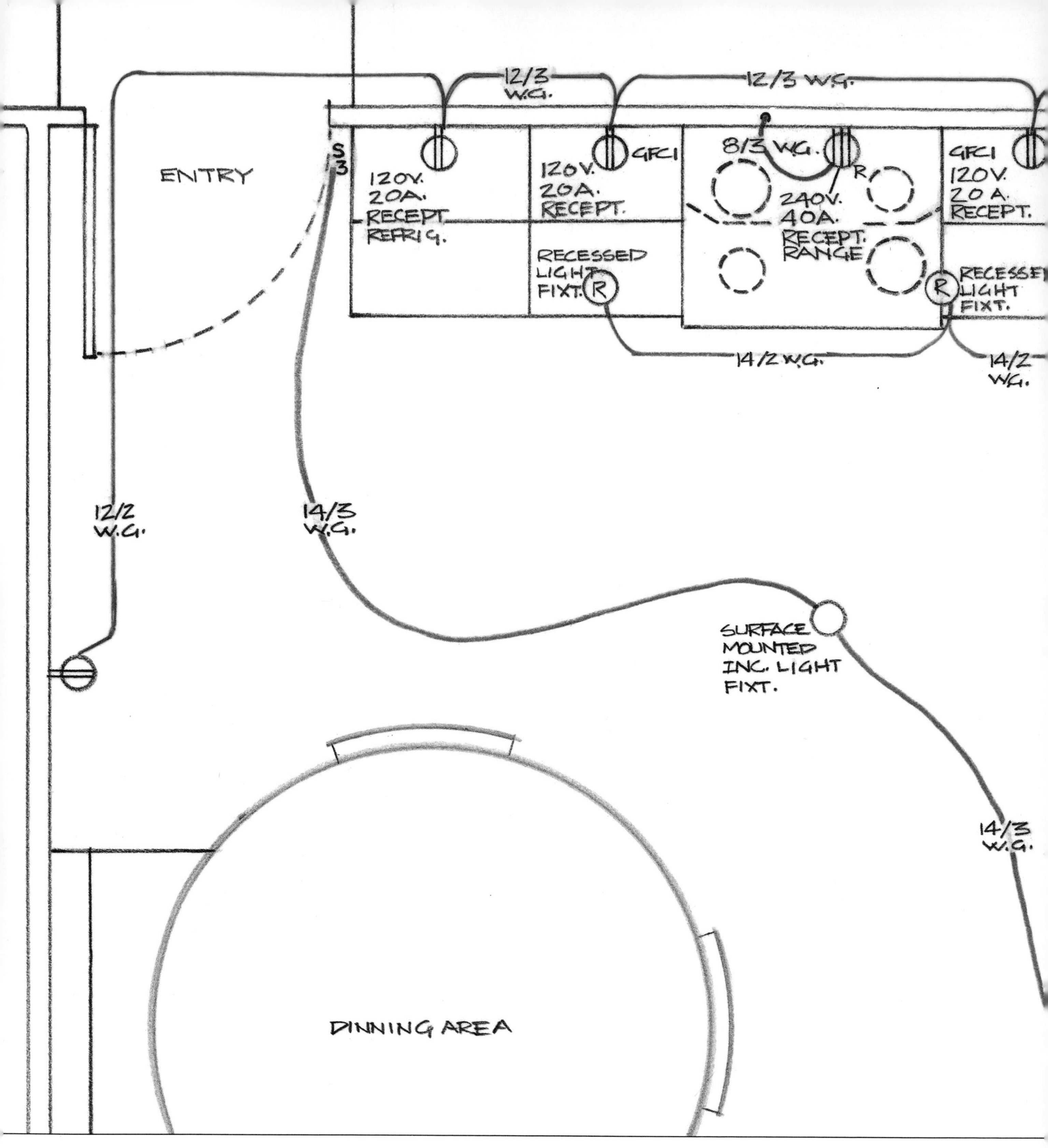

Circuits #1 & #2: Two 20-amp, 120-volt small-appliance circuits wired with one cable. All general-use receptacles must be on these circuits and they must be GFCI units. Includes: 7 GFCI receptacles rated for 20 amps, 5 electrical boxes that are 4" x 4", and 12/3 cable. One GFCI shares a double-gang box with circuit (#5), and another GFCI shares a triple-gang box with circuit (#7).

Circuit #3: A 50-amp, 120/240-volt dedicated circuit for the range. Includes: a 4" × 4" box; a 120/240-volt, 50-amp range receptacle; and 6/3 NM cable.

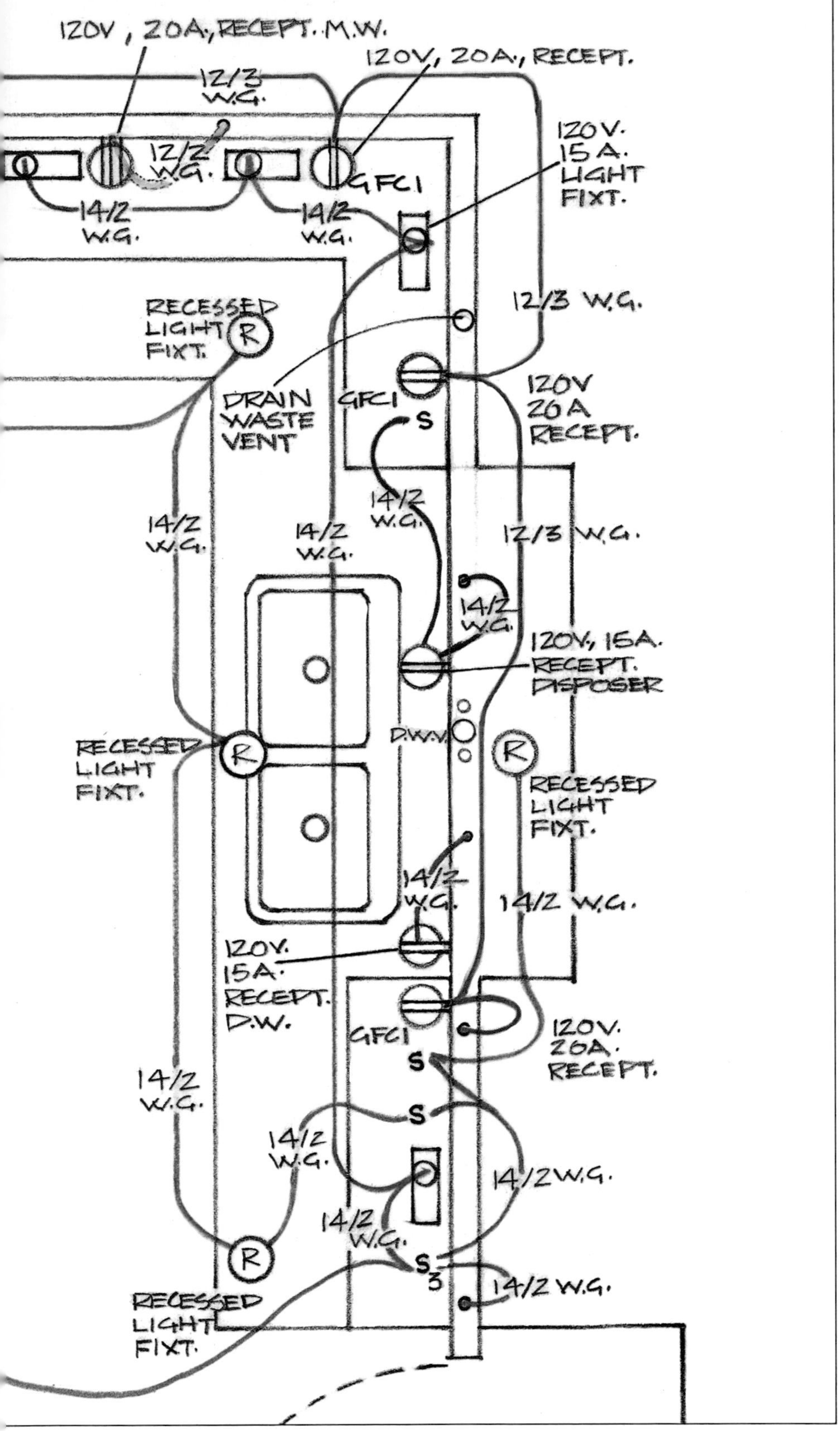

Wiring a Remodeled Kitchen:
Diagram View

This diagram view shows the layout of seven circuits and the location of the switches, receptacles, lights, and other fixtures in the remodeled kitchen featured in this section. The size and number of circuits, and the specific features included are based on the needs of this 170-sq. ft. space. No two remodeled kitchens are exactly alike, so create your own wiring diagram to guide you through your wiring project.

Circuit #7: A 15-amp, 120-volt basic lighting circuit serving all of the lighting needs in the kitchen. Includes: 2 single-pole switches, 2 three-way switches, single-gang box, 4" × 4" box, triple-gang box (shared with one of the GFCI receptacles from the small-appliance circuits), plastic light fixture box with brace, ceiling light fixture, 4 fluorescent under-cabinet light fixtures, 6 recessed light fixtures, 14/2 and 14/3 cable.

Circuit #6: A 15-amp, 120-volt dedicated circuit for the dishwasher. Includes: a 15-amp duplex receptacle, one single-gang box, and 14/2 cable.

Circuit #4: A 20-amp, 120-volt dedicated circuit for the microwave. Includes: a 20-amp duplex receptacle, a single-gang box, and 12/2 NM cable.

Circuit #5: A 15-amp, 120-volt dedicated circuit for the food disposer. Includes: a 15-amp duplex receptacle, a single-pole switch (installed in a double-gang box with a GFCI receptacle from the small-appliance circuits), one single-gang box, and 14/2 cable.

Code requires receptacles above countertops to be no more than 4 ft. apart. Put receptacles closer together in areas where many appliances will be used. Any section of countertop that is wider than 12" must have a receptacle located above it. (Countertop spaces separated by items such as range tops, sinks, and refrigerators are considered separate sections.) All accessible receptacles in kitchens (and bathrooms) must be GFCI protected. On walls without countertops, receptacles should be no more than 12 ft. apart.

Wiring a Remodeled Kitchen

1: Plan the Circuits

A kitchen generally uses the most power in the home because it contains many light fixtures and appliances. Where these items are located depends upon your needs. Make sure plenty of light and enough receptacles will be in the main work areas of your kitchen. Try to anticipate future needs: for example, install a range receptacle when remodeling, even if you currently have a gas range. It is difficult and expensive to make changes later. See pages 6 to 11 for more information on planning circuits.

Contact your local Building and Electrical Code offices before you begin planning. They may have requirements that differ from the National Electrical Code. Remember that the Code contains minimum requirements primarily concerning safety, not convenience or need. Work with the inspectors to create a safe plan that also meets your needs.

To help locate receptacles, plan carefully where cabinets and appliances will be in the finished project. Appliances installed within cabinets, such as microwaves or food disposers, must have their receptacles positioned according to manufacturer's instructions. Put at least one receptacle at table height in the dining areas for convenience in operating a small appliance.

The ceiling fixture should be centered in the kitchen ceiling. Or, if your kitchen contains a dining area or breakfast nook, you may want to center the light fixture over the table. Locate recessed light fixtures and under-cabinet task lights where they will best illuminate main work areas.

Before drawing diagrams and applying for a permit, evaluate your existing service and make sure it provides enough power to supply the new circuits you are planning to add. If it will not, contact a licensed electrician to upgrade your service before beginning your work.

Bring the wiring plan and materials list to the inspector's office when applying for the permit. If the inspector suggests improvements to your plan, such as using switches with grounding screws, follow his advice. He can save you time and money.

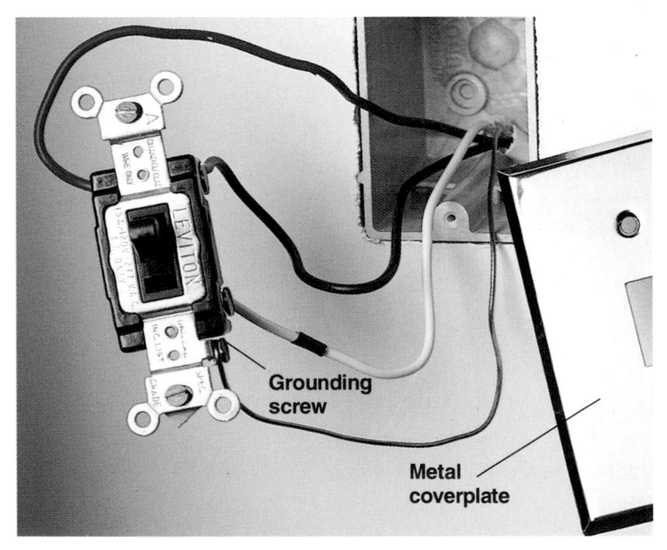

A switch with a grounding screw may be required by inspectors in kitchens and baths. Code requires them when metal coverplates are used with plastic boxes.

Tips for Planning Kitchen Circuits

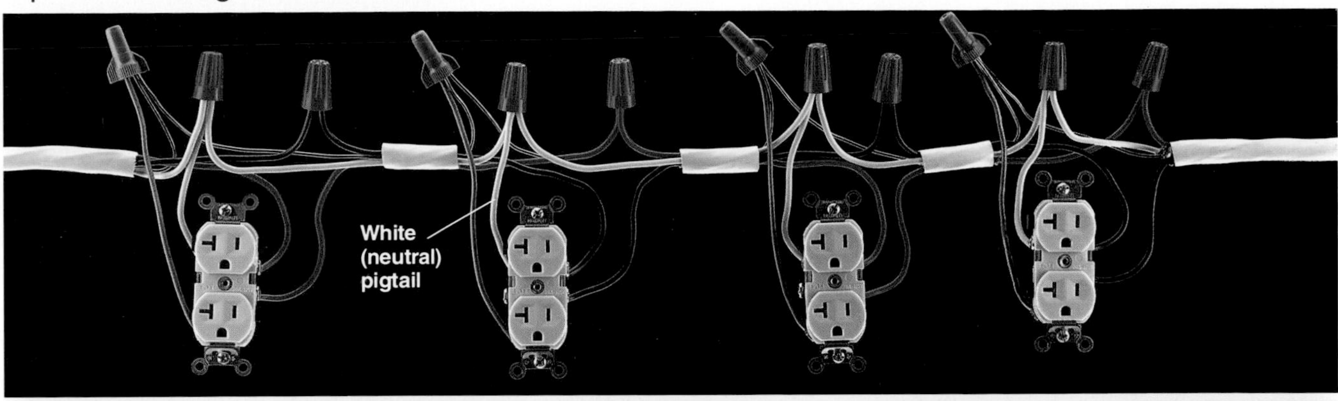

Two 20-amp small-appliance circuits can be wired with one 12/3 cable supplying power to both circuits (top), rather than using separate 12/2 cables for each circuit (bottom), to save time and money. Because these circuits must be GFCI protected, either place a GFCI receptacle first in each circuit (the remaining 20-amp duplex units are connected through the LOAD terminals on the GFCI) or use a GFCI receptacle at each location. In 12/3 cable, the black wire supplies power to one circuit for alternate receptacles (the first, third, etc.), the red wire supplies power for the second circuit to the remaining receptacles. The white wire is the neutral for both circuits. For safety, it must be attached with a pigtail to each receptacle, instead of being connected directly to the terminal. These circuits must contain all general-use receptacles in the kitchen, pantry, breakfast area or dining room. No lighting outlets or receptacles from any other rooms can be connected to them.

Work areas at sink and range should be well lighted for convenience and safety. Install switch-controlled lights over these areas.

Ranges require a dedicated 40- or 50-amp 120/240-volt circuit (or two circuits for separate oven and countertop units). Even if you do not have an electric range, it is a good idea to install the circuit when remodeling.

Dishwashers and food disposers require dedicated 15-amp, 120-volt circuits in most local Codes. Some inspectors will allow these appliances to share one circuit.

Heights of electrical boxes in a kitchen vary depending upon their use. In the kitchen project shown here the centers of the boxes above the countertop are 45" above the floor, in the center of 18" backsplashes that extend from the countertop to the cabinets. All boxes for wall switches also are installed at this height. The center of the box for the microwave receptacle is 72" off the floor, where it will fit between the cabinets. The centers of the boxes for the range and food disposer receptacles are 12" off the floor, but the center of the box for the dishwasher receptacle is 6" off the floor, next to the space the appliance will occupy.

Wiring a Remodeled Kitchen

2: Install Boxes & Cables

After the inspector issues you a work permit, you can begin installing electrical boxes for switches, receptacles, and fixtures. Install all boxes and frames for recessed fixtures such as vent fans and recessed lights before cutting and installing any cable. However, some surface-mounted fixtures, such as under-cabinet task lights, have self-contained wire connection boxes. These fixtures are installed after the walls are finished and the cabinets are in place.

First determine locations for the boxes above the countertops (page opposite). After establishing the height for these boxes, install all of the other visible wall boxes at this height. Boxes that will be behind appliances or inside cabinets should be located according to appliance manufacturer's instructions. For example, the receptacle for the dishwasher cannot be installed directly behind the appliance; it is often located in the sink cabinet for easy access.

Always use the largest electrical boxes that are practical for your installation. Using large boxes ensures that you will meet Code regulations concerning box volume, and simplifies making the connections.

After all the boxes and recessed fixtures are installed, you are ready to measure and cut the cables. First install the feeder cables that run from the circuit breaker panel to the first electrical box in each circuit. Then cut and install the remaining cables to complete the circuits.

Tips for Installing Boxes & Cables

Use masking tape to outline the location of all cabinets, large appliances, and other openings. The outlines help you position the electrical boxes accurately. Remember to allow for moldings and other trim.

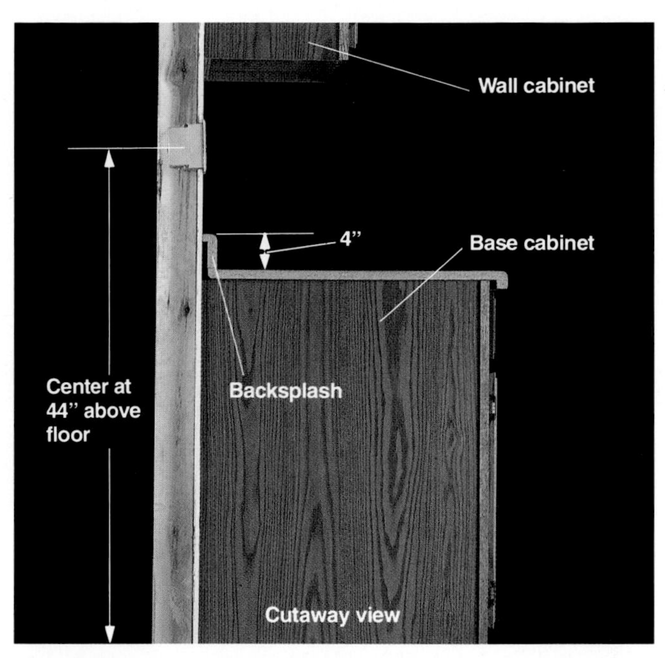

Standard backsplash height is 4"; the center of a box installed above this should be 44" above the floor. If the backsplash is more than 4" high, or the distance between the countertop and the bottom of the cabinet is less than 18", center the box in the space between the countertop and the bottom of the wall cabinet.

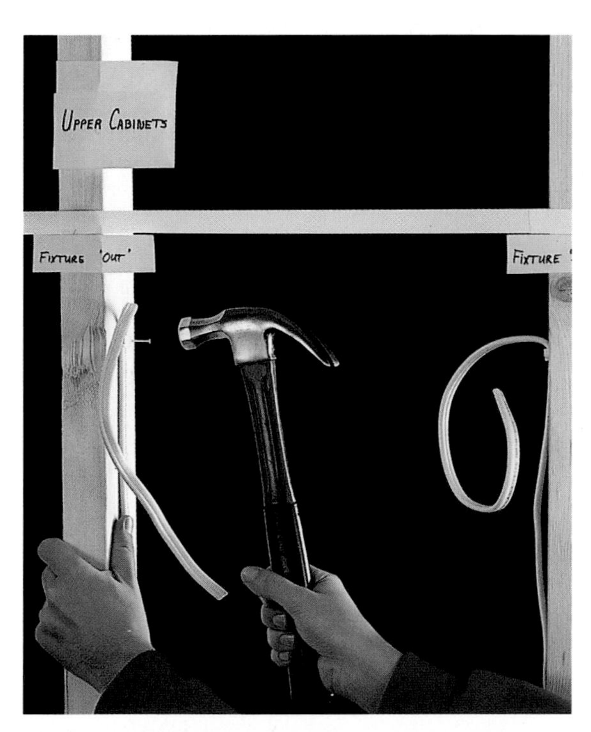

Install cables for an under-cabinet light at positions that will line up with knockouts on the fixture box (which is installed after the walls and cabinets are in place). Cables will be retrieved through ⅝" drilled holes (page 51), so it is important to position the cables accurately.

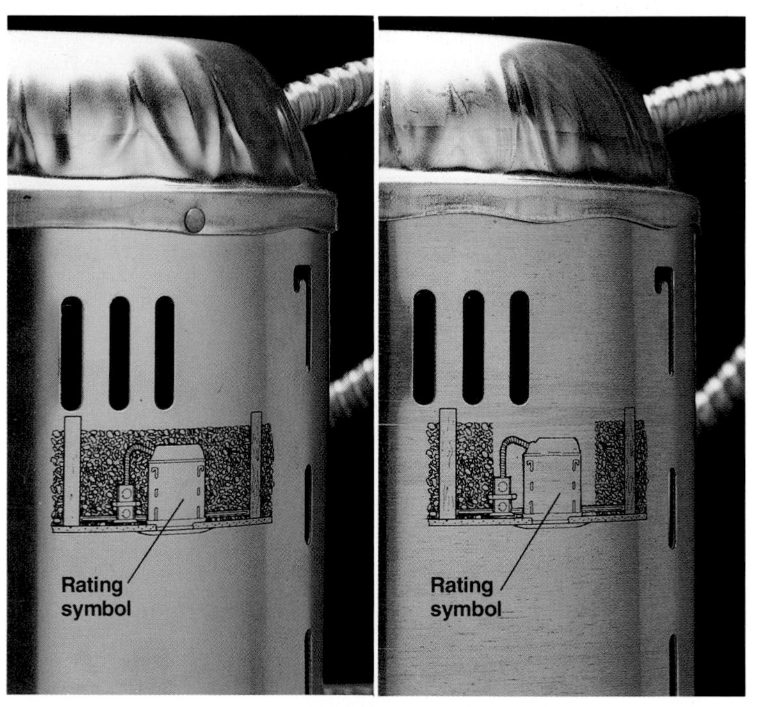

Choose the proper type of recessed light fixture for your project. There are two types of fixtures: those rated for installation within insulation (left), and those which must be kept at least 3" from insulation (right). Self-contained thermal switches shut off power if the unit gets too hot for its rating. A recessed light fixture must be installed at least 1/2" from combustible materials.

How to Mount a Recessed Light Fixture

1 Extend the mounting bars on the recessed fixture to reach the framing members. Adjust the position of the light unit on the mounting bars to locate it properly. Align the bottom edges of the mounting bars with the bottom face of the framing members.

2 Nail or screw the mounting bars to the framing members.

3 Remove the wire connection box cover and open one knockout for each cable entering the box.

4 Install a cable clamp for each open knockout, and tighten locknut, using a screwdriver to drive the lugs.

How to Install the Feeder Cable

1 Drill access holes through the sill plate where the feeder cables will enter from the circuit breaker panel. Choose spots that offer easy access to the circuit breaker panel as well as to the first electrical box on the circuit.

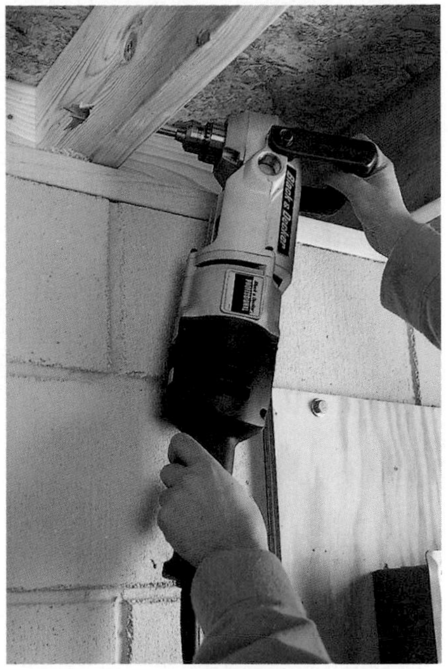

2 Drill 5/8" holes through framing members to allow cables to pass from the circuit breaker panel to access holes. Front edge of hole should be at least 1 1/4" from front edge of framing member.

3 For each circuit, measure and cut enough cable to run from circuit breaker panel, through access hole into the kitchen, to the first electrical box in the circuit. Add at least 2 ft. for the panel and 1 ft. for the box.

4 Anchor the cable with a cable staple within 12" of the panel. Extend cable through and along joists to access hole into kitchen, stapling every 4 ft. where necessary. Keep cable at least 1¼" from front edge of framing members. Thread cable through access hole into kitchen, and on to the first box in the circuit. Continue circuit to rest of boxes.

Arrange for the rough-in inspection before making the final connections.

3: Make Final Connections

Make the final connections for switches, receptacles, and fixtures after the rough-in inspection. First make final connections on recessed fixtures (it is easier to do this before wallboard is installed). Then finish the work on walls and ceiling, install the cabinets, and make the rest of the final connections. Use the photos on the following pages as a guide for making the final connections. The last step is to connect the circuits at the breaker panel. After all connections are made, your work is ready for the final inspection.

Materials You Will Need:

Pigtail wires, wire nuts, black tape.

Circuits #1 & #2

Two 20-amp, 120-volt small-appliance circuits.

- 7 GFCI receptacles
- 20-amp double-pole circuit breaker

Note: In this project, two of the GFCI receptacles are installed in boxes that also contain switches from other circuits (page opposite).

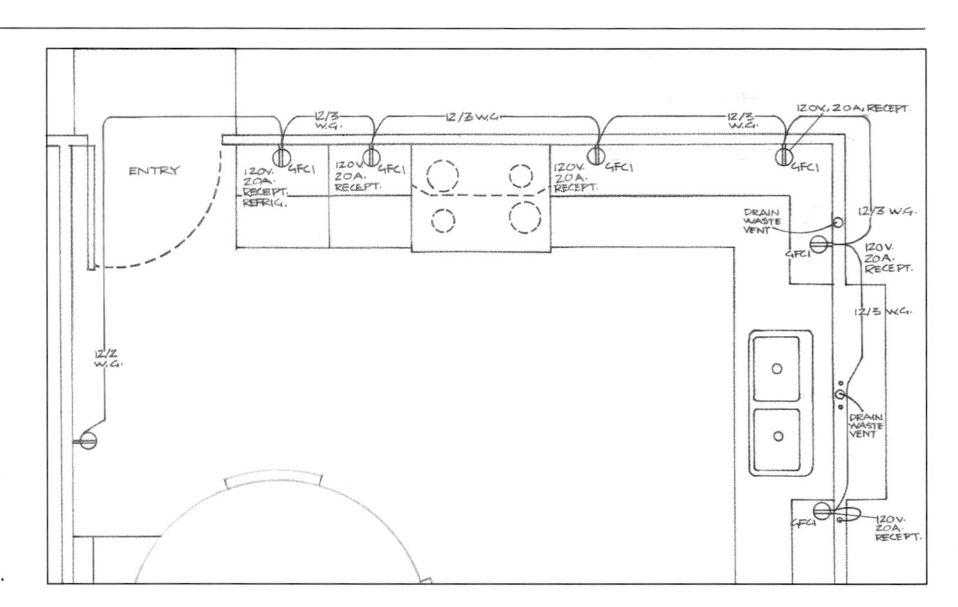

How to Connect Small-appliance Receptacles (that alternate on two 20-amp circuits in one 12/3 cable)

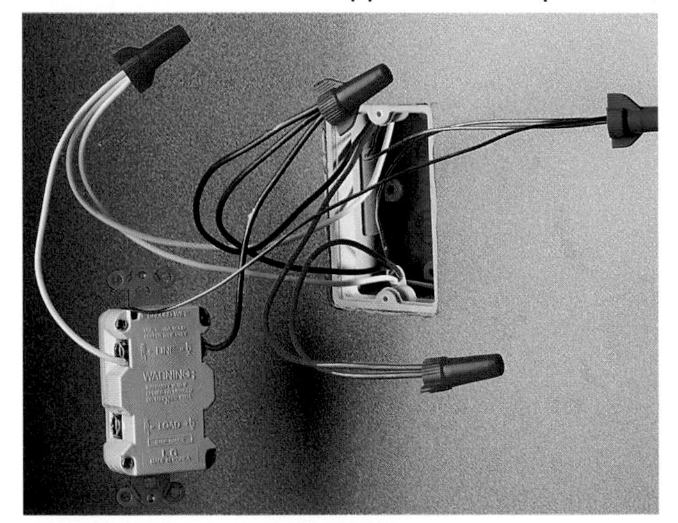

1 At alternate receptacles in the cable run (first, third, etc.), attach a black pigtail to brass screw terminal marked LINE on the receptacle and to black wire from both cables. Connect a white pigtail to a silver screw (LINE) and to both white wires. Connect a grounding pigtail to the grounding screw and to both grounding wires. Connect both red wires together. Tuck wires into box, then attach the receptacle and coverplate.

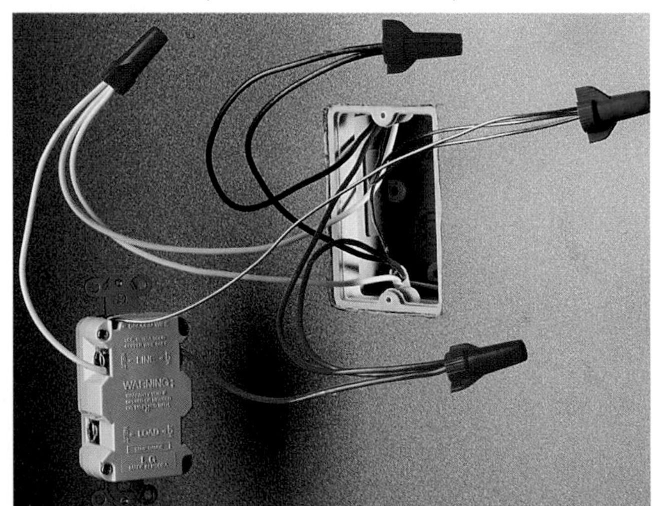

2 At remaining receptacles in the run, attach a red pigtail to a brass screw terminal (LINE) and to red wires from the cables. Attach a white pigtail to a silver screw terminal (LINE) and to both white wires. Connect a grounding pigtail to the grounding screw and to both grounding wires. Connect both black wires together. Tuck wires into box, attach receptacle and coverplate. (See page 41 for optional method of GFCI protection.)

How to Install a GFCI & a Disposer Switch

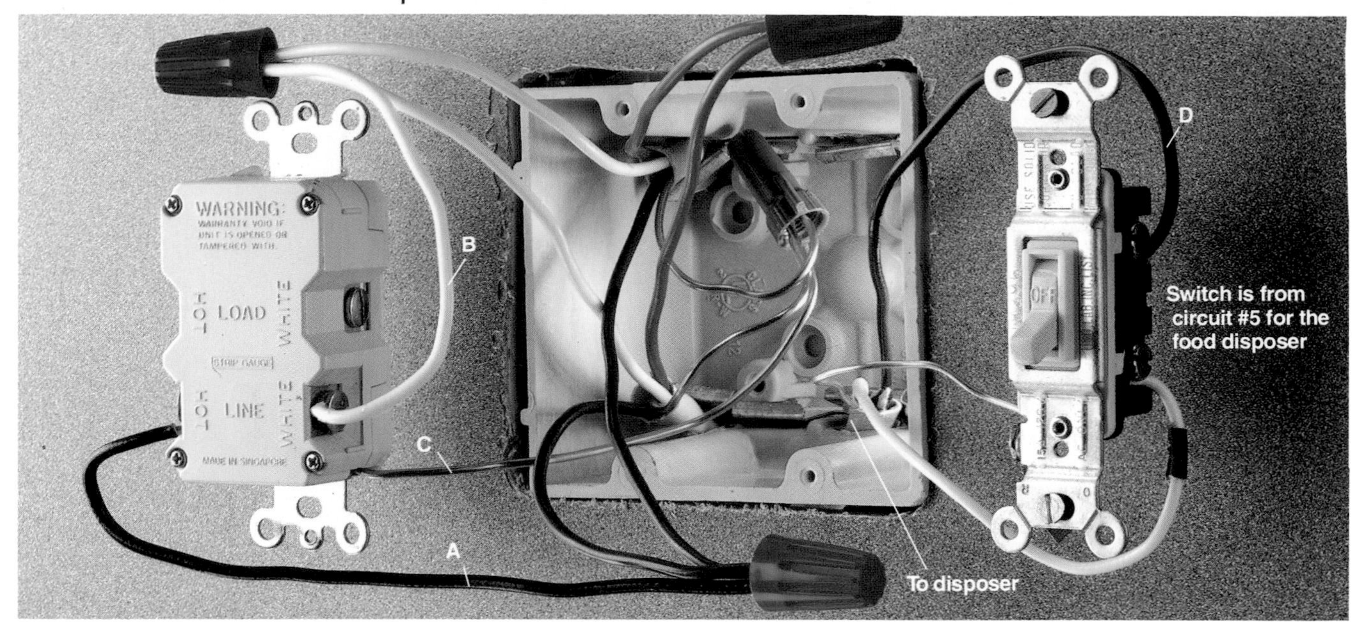

Connect black pigtail (A) to GFCI brass terminal marked LINE, and to black wires from three-wire cables. Attach white pigtail (B) to silver terminal marked LINE, and to white wires from three-wire cables. Attach grounding pigtail (C) to GFCI grounding screw and to grounding wires from three-wire cables. Connect both red wires together. Connect black wire from two-wire cable (D) to one switch terminal. Attach white wire to other terminal and tag it black indicating it is hot. Attach grounding wire to switch grounding screw. Tuck wires into box, attach switch, receptacle, and coverplate.

How to Install a GFCI & Two Switches for Recessed Lights

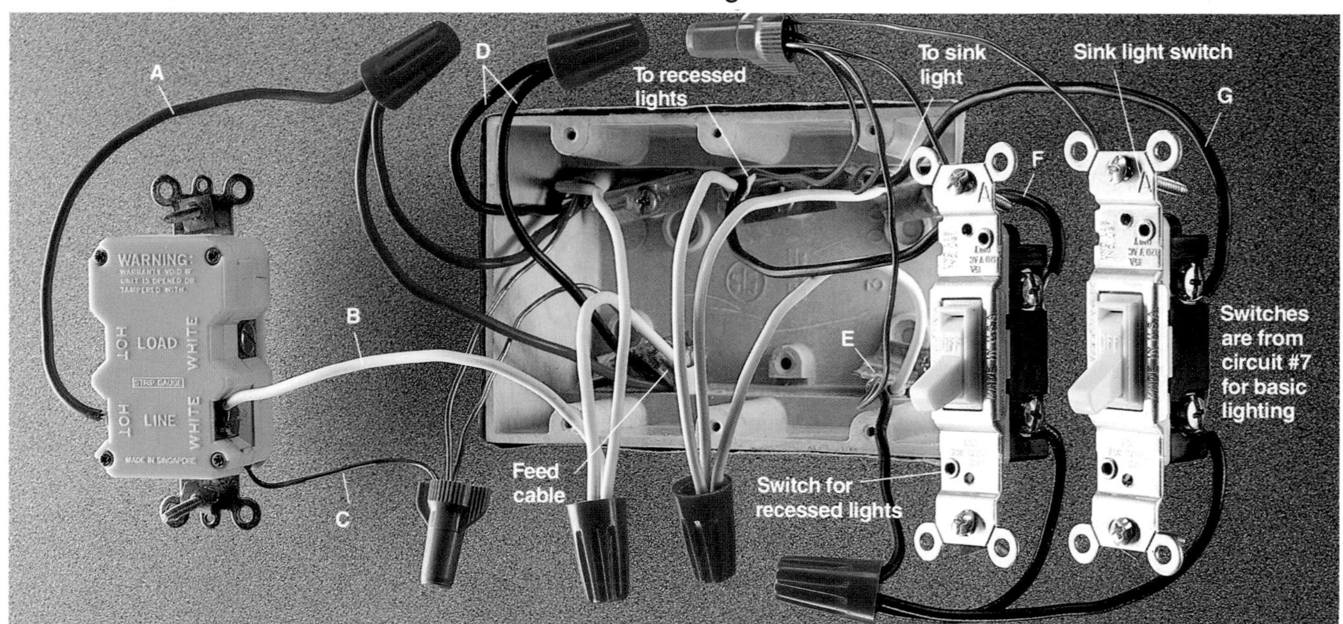

Connect red pigtail (A) to GFCI brass terminal labeled LINE, and to red wires from three-wire cables. Connect white pigtail (B) to silver LINE terminal, and to white wires from three-wire cables. Attach grounding pig tail (C) to grounding screw, and to grounding wires from three-wire cables. Connect black wires from three-wire cables (D) together. Attach a black pigtail to one screw on each switch and to black wire from two-wire feed cable (E). Connect black wire (F) from the two-wire cable leading to recessed lights to remaining screw on the switch for the recessed lights. Connect black wire (G) from two-wire cable leading to sink light to remaining screw on sink light switch. Connect white wires from all two-wire cables together. Connect pigtails to switch grounding screws, and to all grounding wires from two-wire cables. Tuck wires into box, attach switches, receptacles, and coverplate.

Circuit #3

A 50-amp, 120/240-volt circuit serving the range.

- 50-amp receptacle for range
- 50-amp double-pole circuit breaker

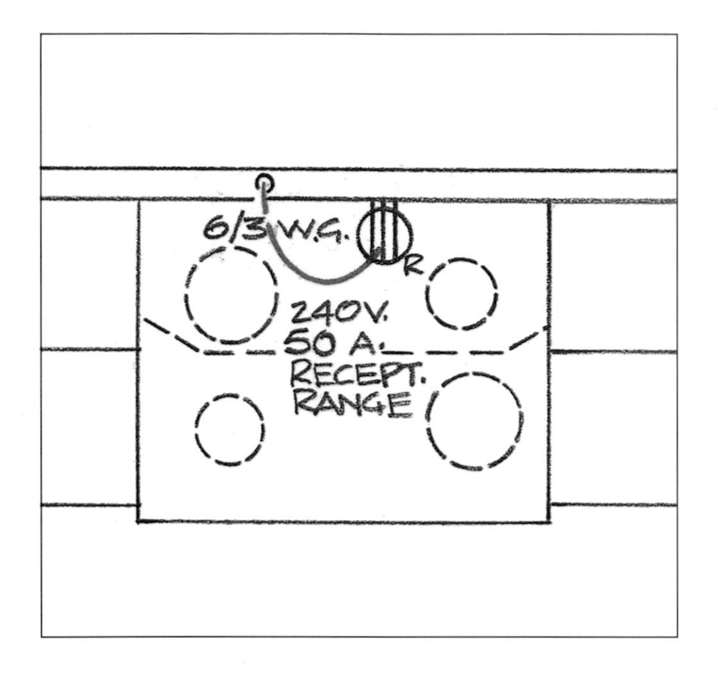

How to Install 120/240 Range Receptacle

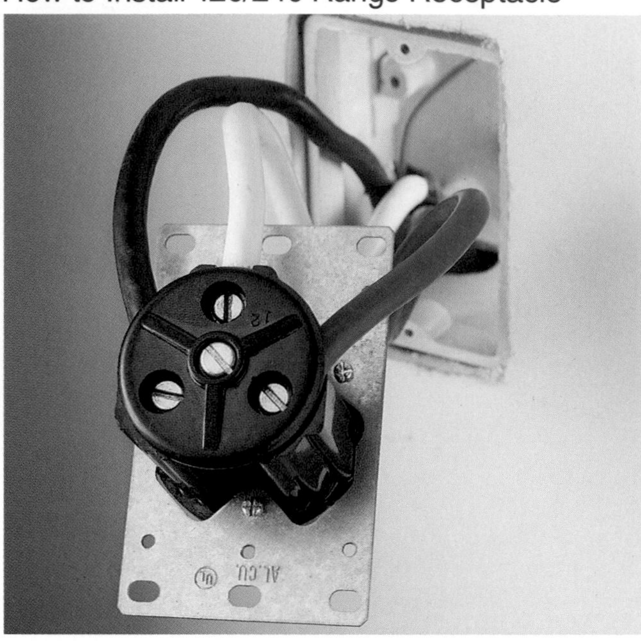

Attach the white wire to the neutral terminal, and the black and red wires to the remaining terminals. The neutral white wire acts as the grounding wire for this circuit, so push the bare copper ground wire from the cable to the back of the box. Tuck the rest of the wires into the box. Attach receptacle and coverplate.

Circuit #4

A 20-amp, 120-volt circuit for the microwave.

- 20-amp duplex receptacle
- 20-amp single-pole circuit breaker

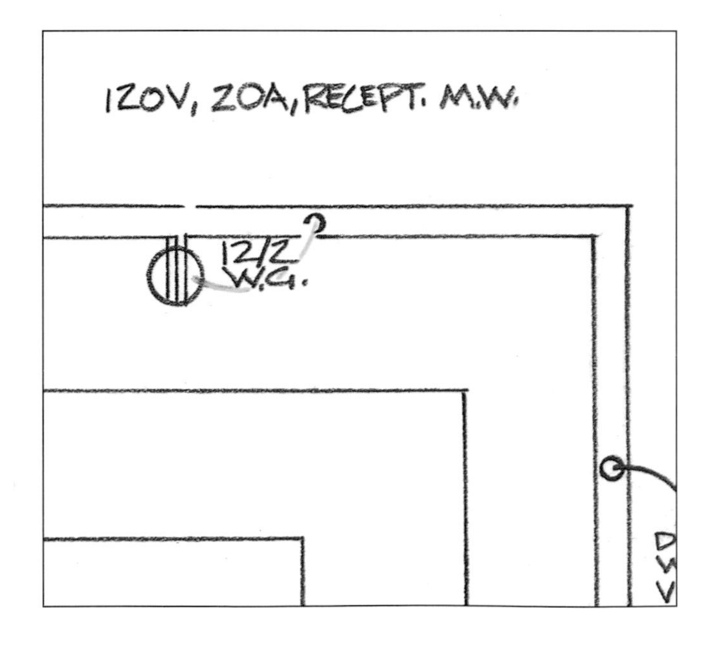

How to Connect Microwave Receptacle

Connect black wire from the cable to a brass screw terminal on the receptacle. Attach the white wire to a silver screw terminal, and the grounding wire to the receptacle's grounding screw. Tuck wires into box, attach the receptacle and the coverplate.

Circuit #5

A 15-amp, 120-volt circuit for the food disposer.

- 15-amp duplex receptacle
- Single-pole switch
- 15-amp single-pole circuit breaker

Note: Final connection of the single-pole switch controlling the disposer is shown on page 47.

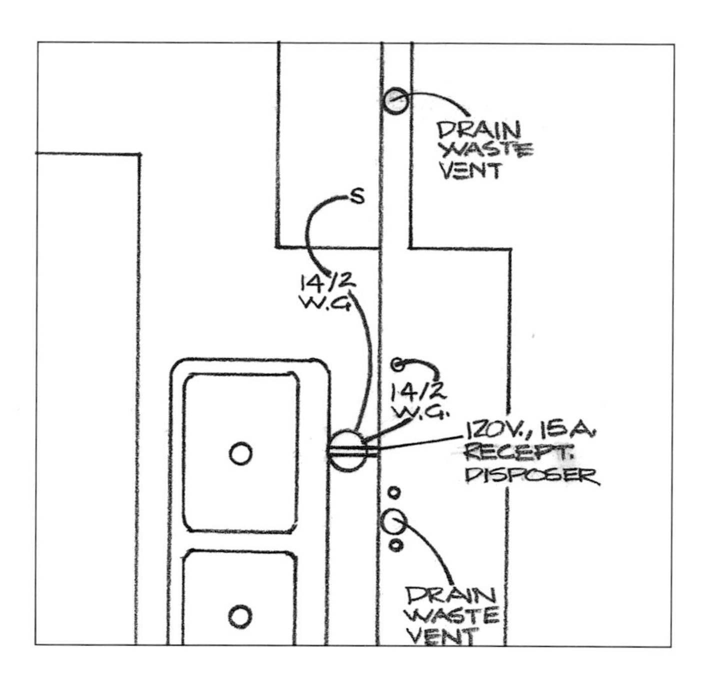

How to Connect Disposer Receptacle

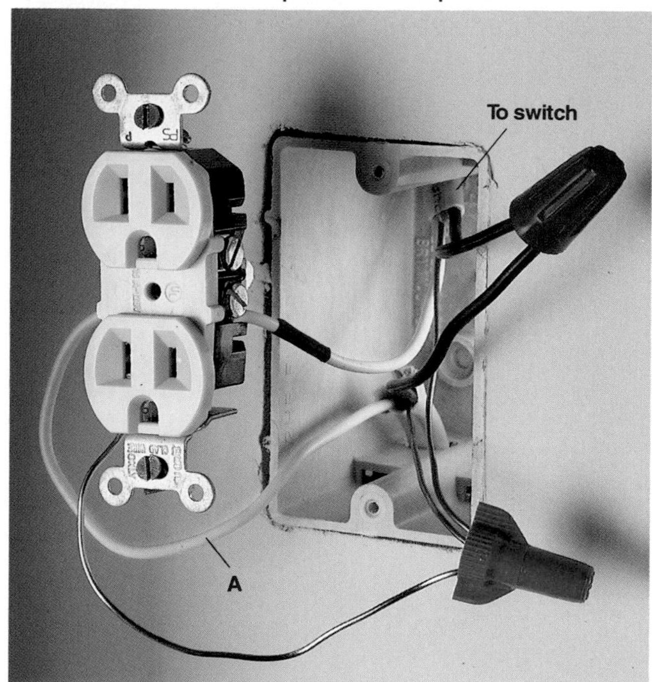

Connect black wires together. Connect white wire from feed cable (A) to silver screw on receptacle. Connect white wire from cable going to the switch to a brass screw terminal on the receptacle, and tag the wire with black indicating it is hot. Attach a grounding pigtail to the grounding screw and to both cable grounding wires. Tuck wires into box, then attach receptacle and coverplate.

Circuit #6

A 15-amp, 120-volt circuit for the dishwasher.

- 15-amp duplex receptacle
- 15-amp single-pole circuit breaker

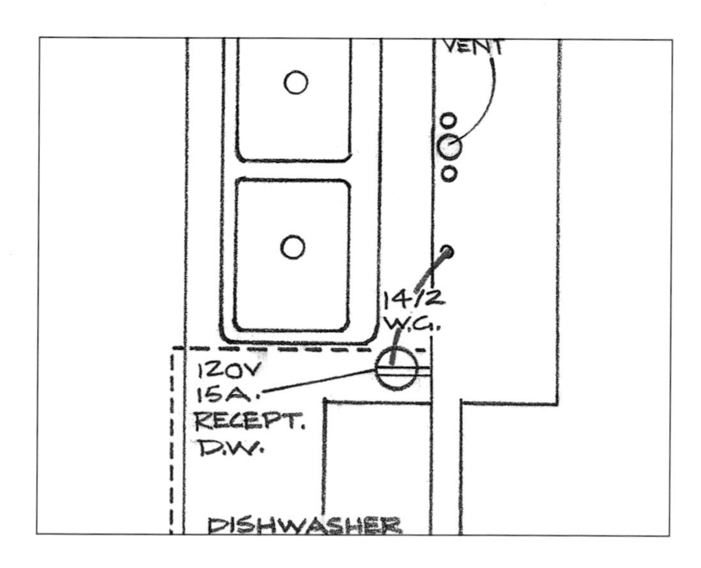

How to Connect Dishwasher Receptacle

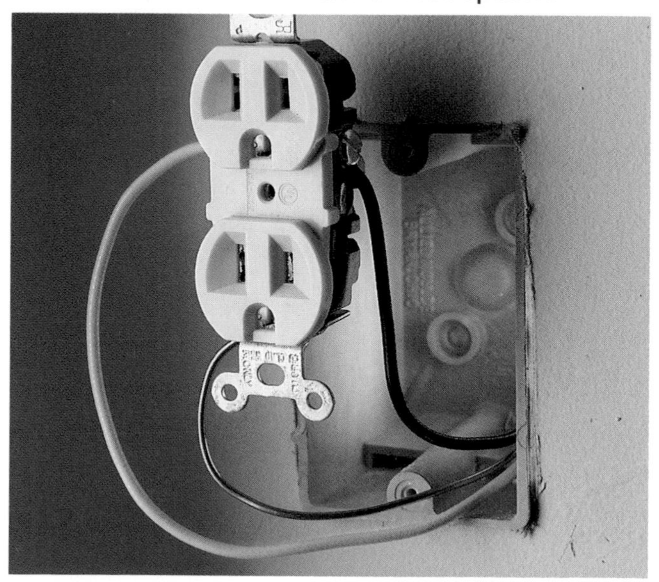

Connect the black wire to a brass screw terminal. Attach the white wire to a silver screw terminal. Connect the grounding wire to the grounding screw. Tuck wires into box, then attach the receptacle and the coverplate.

Circuit #7

A 15-amp basic lighting circuit serving the kitchen.

- 2 three-way switches with grounding screws
- 2 single-pole switches with grounding screws
- Ceiling light fixture
- 6 recessed light fixtures
- 4 fluorescent under-cabinet fixtures
- 15-amp single-pole circuit breaker

Note: Final connections for the single-pole switches are shown on page 47.

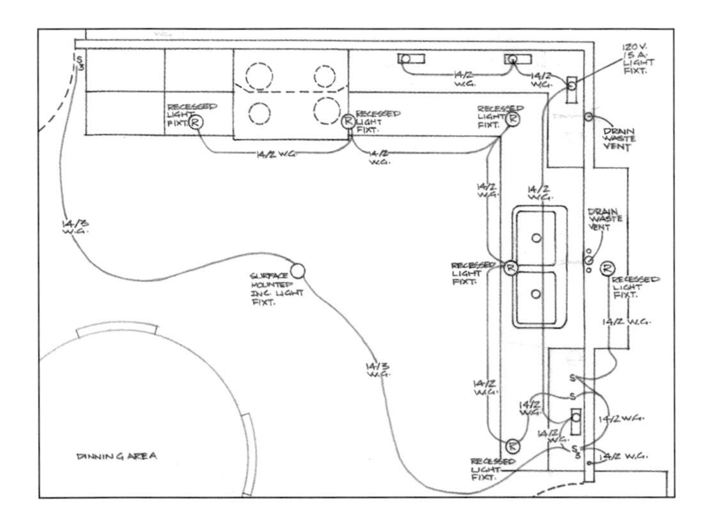

How to Connect Surface-mounted Ceiling Fixture

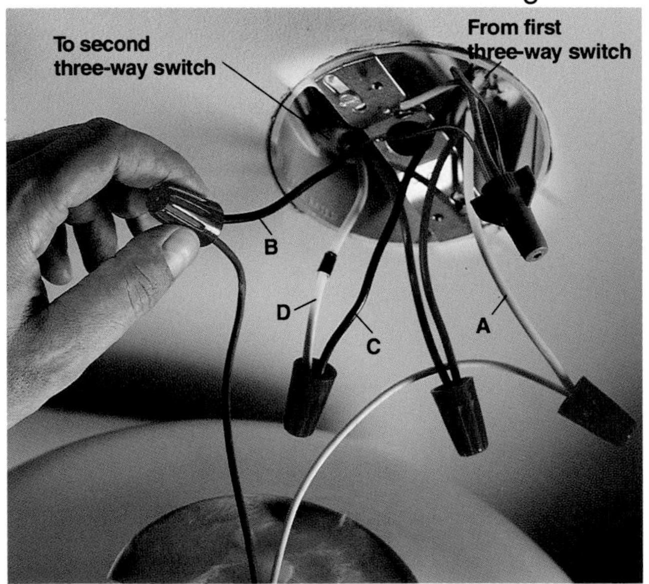

Connect white fixture lead to white wire (A) from first three-way switch. Connect black fixture lead to black wire (B) from second three-way switch. Connect black wire (C) from first switch to white wire (D) from second switch, and tag this white wire with black. Connect red wires from both switches together. Connect all grounding wires together. Mount fixture following manufacturer's instructions.

How to Connect First Three-way Switch

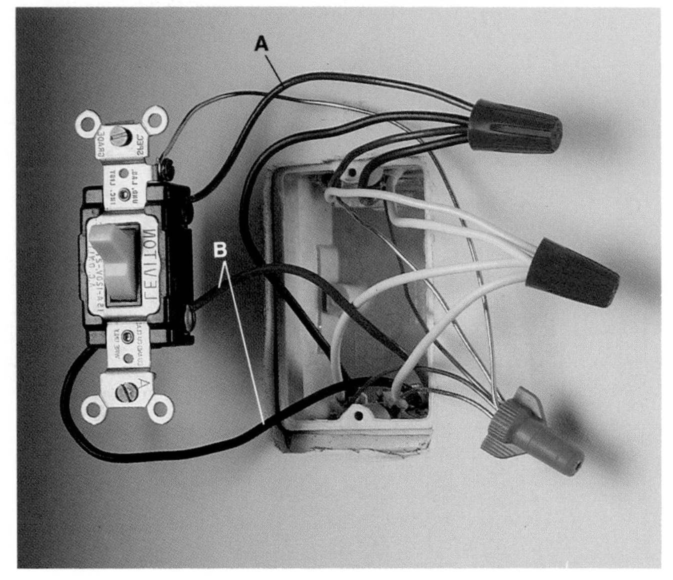

Connect a black pigtail to the common screw on the switch (A) and to the black wires from the two-wire cable. Connect black and red wires from the three-wire cable to traveler terminals (B) on the switch. Connect white wires from all cables entering box together. Attach a grounding pigtail to switch grounding screw and to all grounding wires in box. Tuck wires into box, then attach the switch and the coverplate.

How to Connect Second Three-way Switch

Connect black wire from the cable to the common screw terminal (A). Connect red wire to one traveler screw terminal. Attach the white wire to the other traveler screw terminal and tag it with black, indicating it is hot. Attach the grounding wire to the grounding screw on the switch. Tuck wires into box, then attach the switch and the coverplate.

How to Connect Recessed Light Fixtures

1 Make connections before installing wallboard: the work must be inspected first and access to the junction box is easier. Connect white cable wires to white fixture lead.

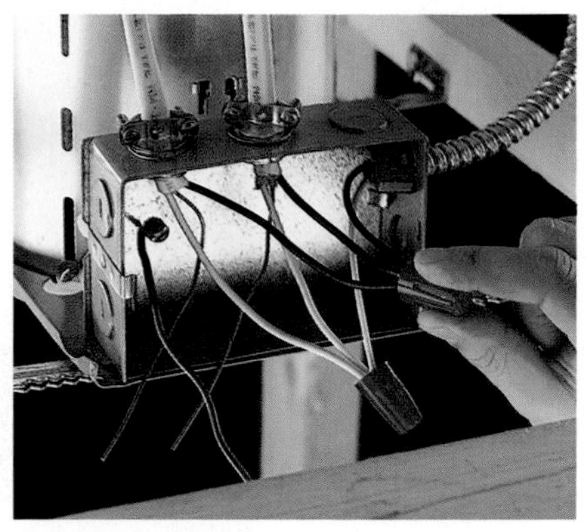

2 Connect black wires to black lead from fixture.

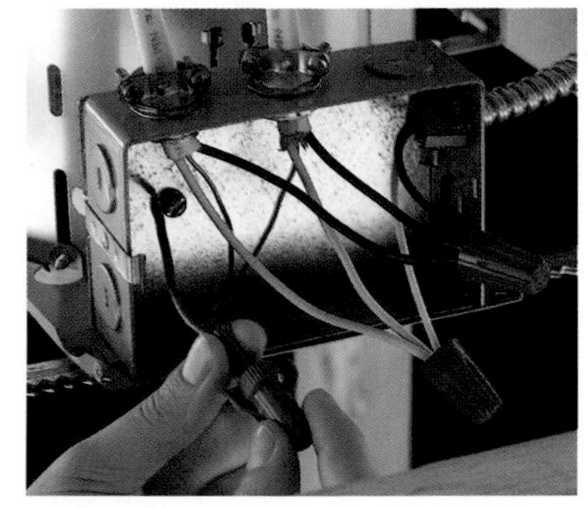

3 Attach a grounding pigtail to the grounding screw on the fixture, then connect all grounding wires. Tuck wires into the junction box, and replace the cover.

How to Connect Under-cabinet Fluorescent Task Light Fixtures

1 Drill ⅝" holes through wall and cabinet at locations that line up with knockouts on the fixture, and retrieve cable ends (page 43).

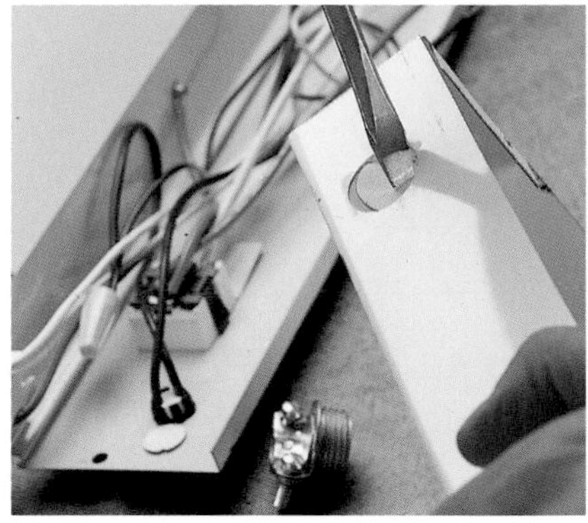

2 Remove access cover on fixture. Open one knockout for each cable that enters fixture box, and install cable clamps.

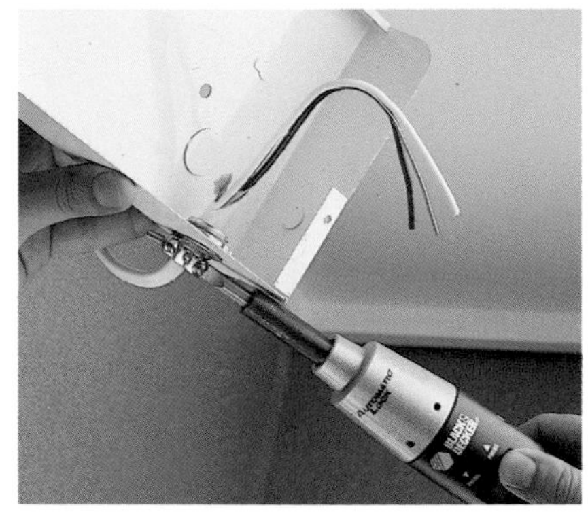

3 Strip 8" of sheathing from each cable end. Insert each end through a cable clamp, leaving 1/4" of sheathing in fixture box.

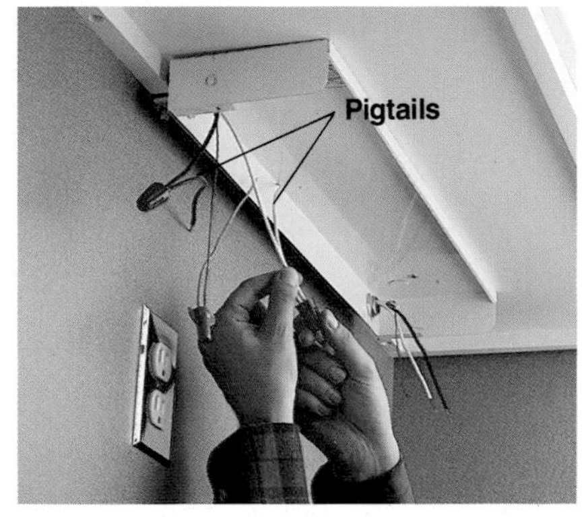

4 Screw fixture box to cabinet. Attach black, white, and green pigtails of THHN/THWN wire to wires from one cable entering box. Pigtails must be long enough to reach the cable at other end of box.

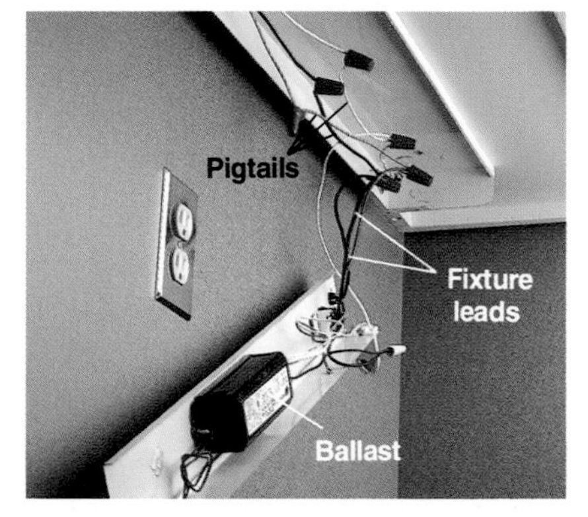

5 Connect black pigtail and circuit wire to black lead from fixture. Connect white pigtail and circuit wire to white lead from fixture. Attach green pigtail and copper circuit wire to green grounding wire attached to the fixture box.

6 Tuck wires into box, and route THHN/THWN pigtails along one side of the ballast. Replace access cover and fixture lens.

Make hookups at circuit breaker panel and arrange for final inspection.

A
B
C

Installing Outdoor Wiring

Adding an outdoor circuit improves the value of your property and lets you enjoy your yard more fully. Doing the work yourself is also a good way to save money. Most outdoor wiring projects require digging underground trenches, and an electrician may charge several hundred dollars for this simple but time-consuming work.

Do not install your own wiring for a hot tub, fountain, or swimming pool. These outdoor water fixtures require special grounding techniques that are best left to an electrician.

In this chapter you learn how to install the following fixtures:

Decorative light fixtures (A) can highlight attractive features of your home and yard, like a deck, ornamental shrubs and trees, and flower gardens. See page 71.

A weatherproof switch (B) lets you control outdoor lights without going indoors. See page 68.

GFCI-protected receptacles (C) let you use electric lawn and garden tools, and provide a place to plug in radios, barbecue rotisseries, and other devices that help make your yard more enjoyable. See page 70.

A manual override switch (D) lets you control a motion-sensor light fixture from inside the house. See page 68.

Five Steps for Installing Outdoor Wiring

1. Plan the circuit (pages 58 to 59).
2. Dig trenches (pages 60 to 61).
3. Install boxes and conduit (pages 62 to 65).
4. Install UF cable (pages 66 to 67).
5. Make final connections (pages 68 to 71).

A motion-sensor light fixture (photos, right) provides inexpensive and effective protection against intruders. It has an infrared eye that triggers the light fixture when a moving object crosses its path. Choose a light fixture with a photo cell (E) that prevents the fixture from triggering in daylight. Look for an adjustable timer (F) that controls how long the light keeps shining after motion stops. Better models have range controls (G) to adjust the sensitivity of the motion-sensor eye. See pages 68 to 69.

Installing Outdoor Wiring:
Cutaway View

The outdoor circuit installation shown on the following pages gives step-by-step instructions for installing a simple outdoor circuit for light fixtures and receptacles. The materials and techniques also can be applied to other outdoor wiring projects, such as running a circuit to a garage workshop, or to a detached shed or gazebo.

Your outdoor wiring probably will be different than the circuit shown in this chapter.

Learn These Techniques for Installing Outdoor Wiring

A: Install weatherproof decorative light fixtures with watertight threaded fittings (page 71).

B: Use rigid metal or IMC conduit with threaded compression fittings (pages 64 to 65) to protect exposed wires and cables.

C: Install a cast-aluminum switch box (page 67) to hold an outdoor switch. The box has a watertight coverplate with toggle lever built into it.

D: Use weatherproof receptacle boxes made of cast aluminum with sealed coverplates and threaded fittings to hold outdoor receptacles (pages 64 to 65).

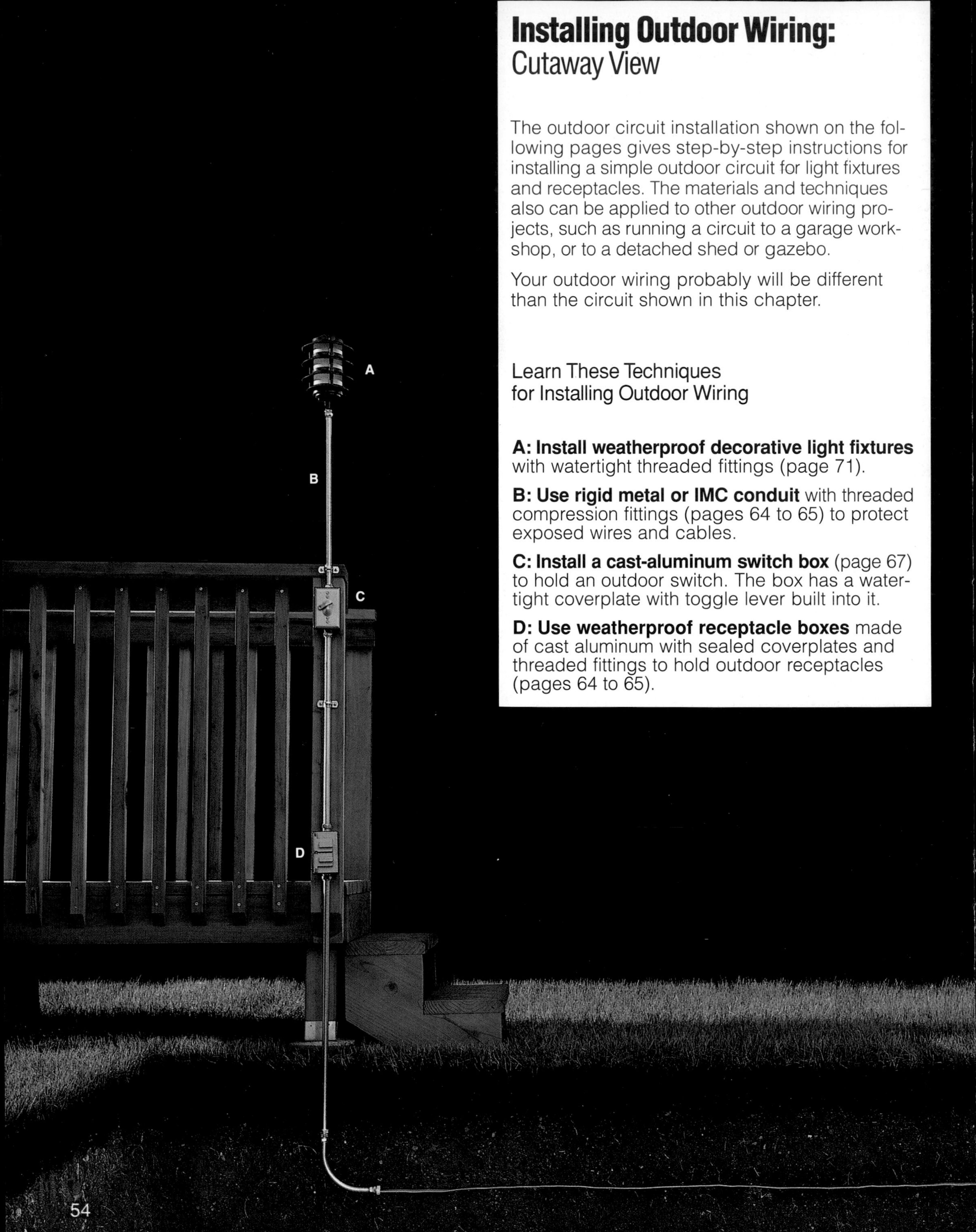

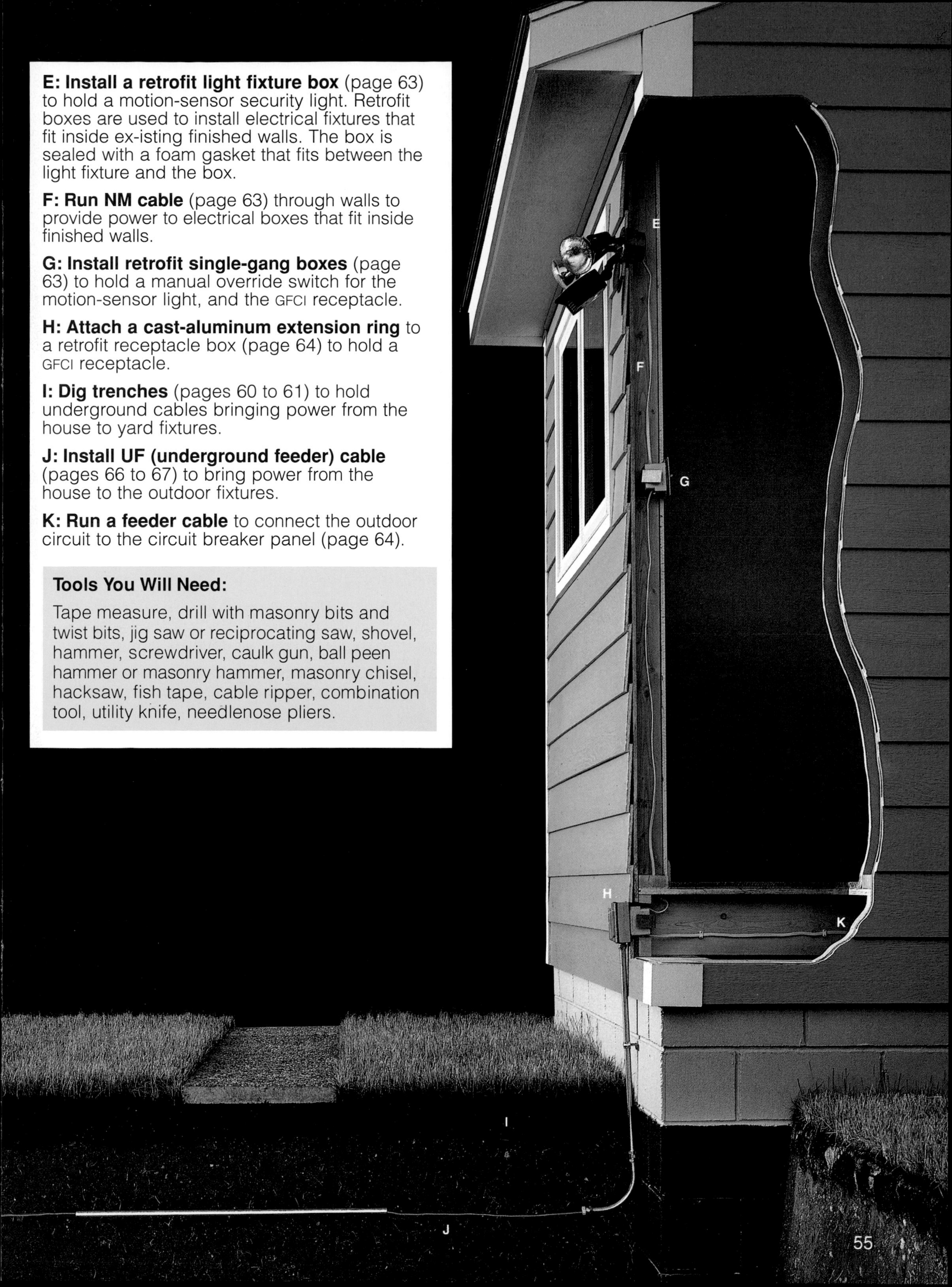

E: Install a retrofit light fixture box (page 63) to hold a motion-sensor security light. Retrofit boxes are used to install electrical fixtures that fit inside ex-isting finished walls. The box is sealed with a foam gasket that fits between the light fixture and the box.

F: Run NM cable (page 63) through walls to provide power to electrical boxes that fit inside finished walls.

G: Install retrofit single-gang boxes (page 63) to hold a manual override switch for the motion-sensor light, and the GFCI receptacle.

H: Attach a cast-aluminum extension ring to a retrofit receptacle box (page 64) to hold a GFCI receptacle.

I: Dig trenches (pages 60 to 61) to hold underground cables bringing power from the house to yard fixtures.

J: Install UF (underground feeder) cable (pages 66 to 67) to bring power from the house to the outdoor fixtures.

K: Run a feeder cable to connect the outdoor circuit to the circuit breaker panel (page 64).

Tools You Will Need:

Tape measure, drill with masonry bits and twist bits, jig saw or reciprocating saw, shovel, hammer, screwdriver, caulk gun, ball peen hammer or masonry hammer, masonry chisel, hacksaw, fish tape, cable ripper, combination tool, utility knife, needlenose pliers.

Installing Outdoor Wiring: Diagram View

This diagram view shows the layout of the outdoor wiring project featured on these pages. It includes the location of the switches, receptacles, light fixtures, and cable runs you will learn how to install in this chapter. The layout of your yard and the location of obstacles will determine the best locations for lights, receptacles, and underground cable runs. The wiring

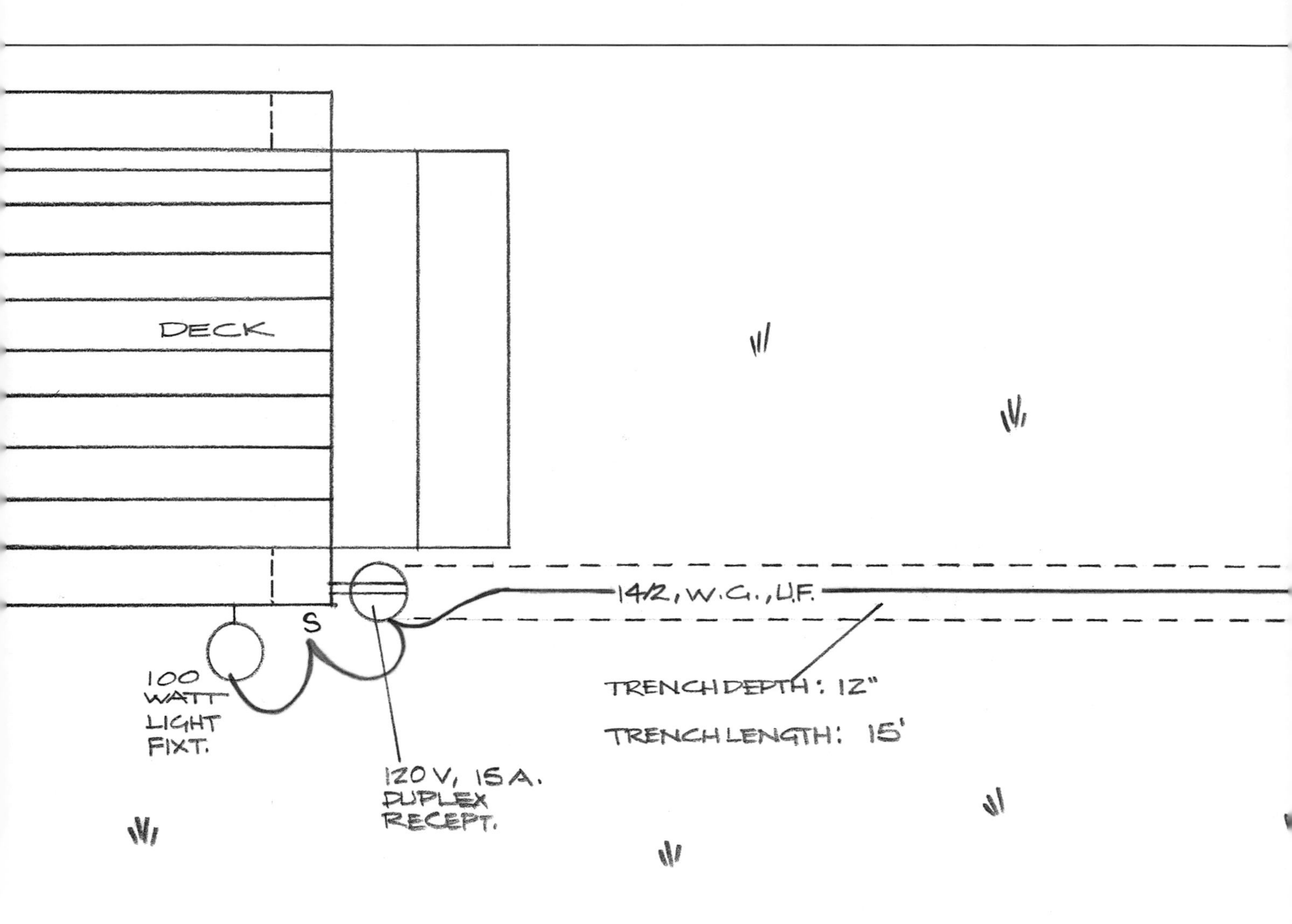

Yard is drawn to scale, with the lengths of trenches and cable runs clearly labeled.

Decorative light fixture is positioned to highlight the deck. Decorative fixtures should be used sparingly, to provide accent only to favorite features of your yard, such as flower beds, ornamental trees, or a patio.

Outdoor receptacle is positioned on the deck post, where it is accessible yet unobtrusive. Another good location for a receptacle is between shrubs.

diagram for your own project may differ greatly from the one shown here, but the techniques shown on the following pages will apply to any outdoor wiring project.

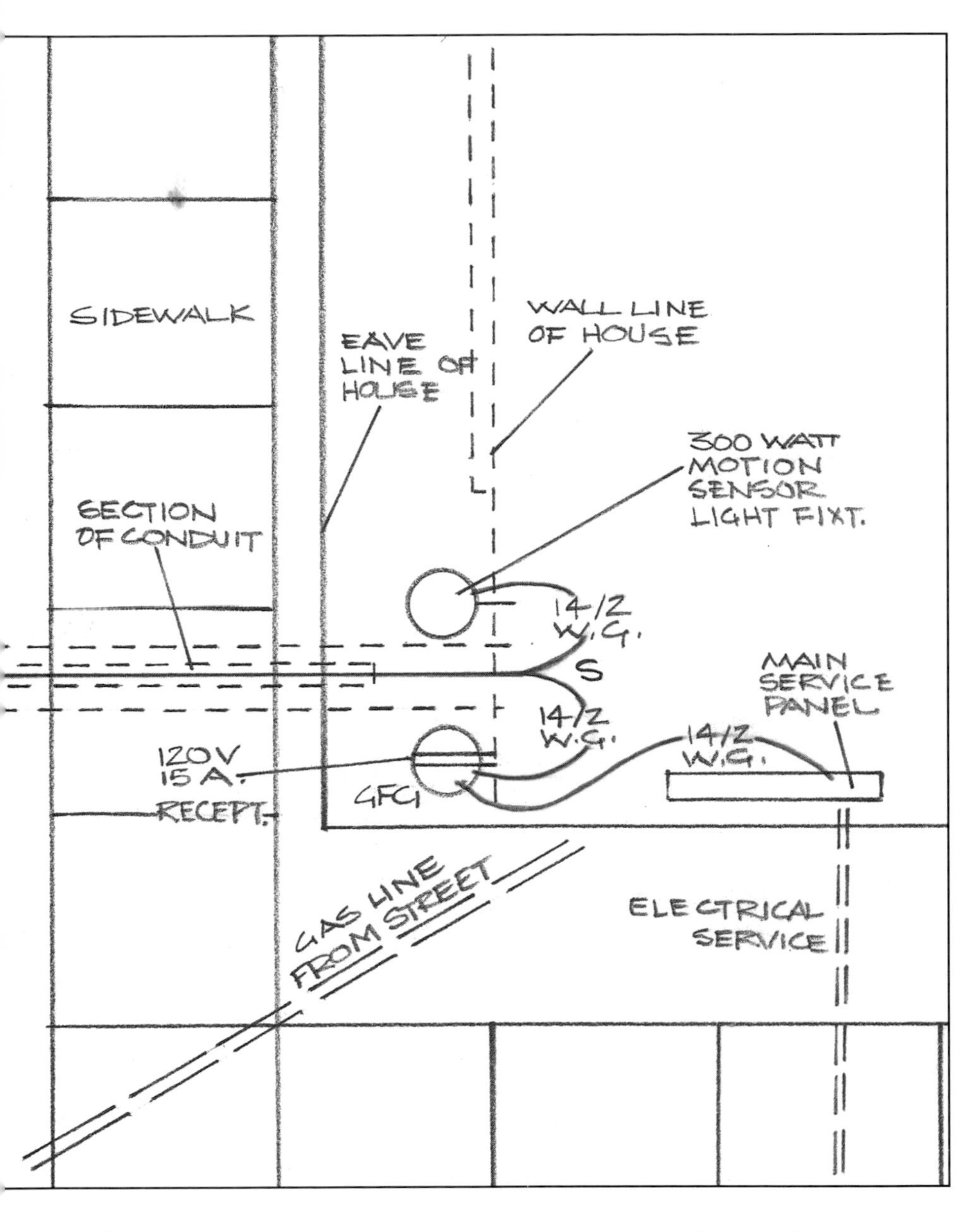

Motion-sensor security light is positioned so it has a good "view" of entryways to the yard and home, and is aimed so it will not shine into neighboring yards.

Manual override switch for motion-sensor light is installed at a convenient indoor location. Override switches are usually mounted near a door or window.

Entry point for circuit is chosen so there is easy access to the circuit breaker panel. Basement rim joists or garage walls make good entry points for an outdoor circuit.

Yard obstacles, like sidewalks and underground gas and electrical lines, are clearly marked as an aid to laying out cable runs.

Underground cables are laid out from the house to the outdoor fixtures by the shortest route possible to reduce the length of trenches.

GFCI **receptacle** is positioned near the start of the cable run, and is wired to protect all wires to the end of the circuit.

Check for underground utilities when planning trenches for underground cable runs. Avoid lawn sprinkler pipes; and consult your electric utility office, phone company, gas and water department, and cable television vendor for the exact locations of underground utility lines. Many utility companies send field representatives to show homeowners how to avoid dangerous underground hazards.

Choosing Cable Sizes for an Outdoor Circuit

Circuit Length		
Less than 50 ft.	50 ft. or more	Circuit size
14-gauge	12-gauge	15-amp
12-gauge	10-gauge	20-amp

Consider the circuit length when choosing cable sizes for an outdoor circuit. In very long circuits, normal wire resistance leads to a substantial drop in voltage. If your outdoor circuit extends more than 50 ft., use larger-gauge wire to reduce the voltage drop. For example, a 15-amp circuit that extends more than 50 ft. should be wired with 12-gauge wire instead of 14-gauge. A 20-amp circuit longer than 50 ft. should be wired with 10-gauge cable.

1: Plan the Circuit

As you begin planning an outdoor circuit, visit your electrical inspector to learn about local Code requirements for outdoor wiring. The techniques for installing outdoor circuits are much the same as for installing indoor wiring. However, because outdoor wiring is exposed to the elements, it requires the use of special weatherproof materials, including UF cable, rigid metal or schedule 40 PVC plastic conduit, and weatherproof electrical boxes and fittings.

The National Electrical Code (NEC) gives minimum standards for outdoor wiring materials, but because climate and soil conditions vary from region to region, your local Building and Electrical Codes may have more restrictive requirements. For example, some regions require that all underground cables be protected with conduit, even though the National Electrical Code allows UF cable to be buried without protection at the proper depths (page opposite).

For most homes, an outdoor circuit is a modest power user. Adding a new 15-amp, 120-volt circuit provides enough power for most outdoor electrical needs. However, if your circuit will include more than three large light fixtures (each rated for 300 watts or more) or more than four receptacles, plan to install a 20-amp, 120-volt circuit. Or, if your outdoor circuit will supply power to heating appliances or large workshop tools in a detached garage, you may require several 120-volt and 240-volt circuits.

Before drawing wiring plans and applying for a work permit, evaluate electrical loads (pages 6 to 7) to make sure the main service provides enough amps to support the added demand of the new wiring.

A typical outdoor circuit takes one or two weekends to install, but if your layout requires very long underground cables, allow yourself more time for digging trenches, or arrange to have extra help. Also make sure to allow time for the required inspection visits when planning your wiring project. See pages 6 to 11 for more information on planning a wiring project.

Tips for Planning an Outdoor Circuit

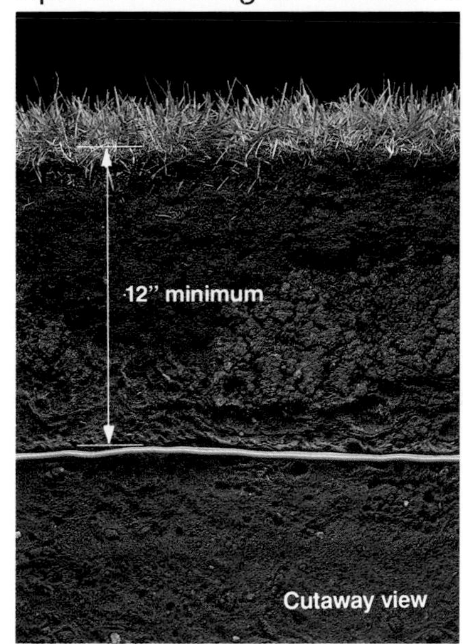

Bury UF cables 12" deep if the wires are protected by a GFCI and the circuit is no larger than 20 amps. Bury cable at least 18" deep if the circuit is not protected by a GFCI, or if it is larger than 20 amps.

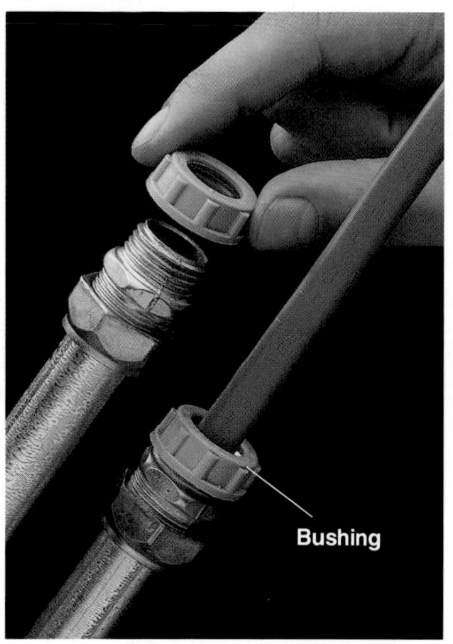

Protect cable entering conduit by attaching a plastic bushing to the open end of the conduit. The bushing prevents sharp metal edges from damaging the vinyl sheathing on the cable.

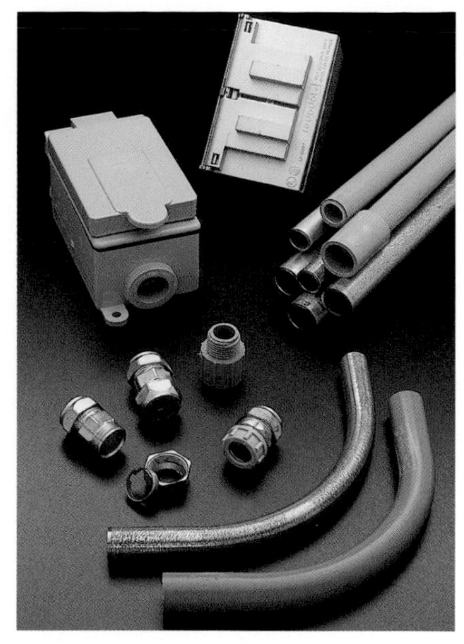

Protect exposed wiring above ground level with rigid conduit and weatherproof electrical boxes and coverplates. Check your local Code restrictions: some regions allow the use of either rigid metal conduit or schedule 40 PVC plastic conduit and electrical boxes, while other regions allow only metal.

Prevent shock by making sure all outdoor receptacles are protected by GFCIs. A single GFCI receptacle can be wired to protect other fixtures on the circuit. Outdoor receptacles should be at least 1 ft. above ground level, and enclosed in weatherproof electrical boxes with watertight covers.

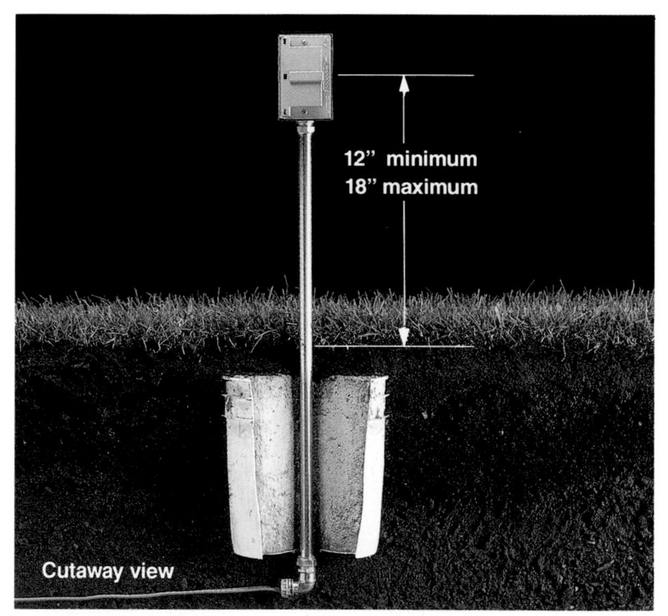

Anchor freestanding receptacles that are not attached to a structure by embedding the rigid metal conduit or schedule 40 PVC plastic conduit in a concrete footing. One way to do this is by running conduit through a plastic bucket, then filling the bucket with concrete. Freestanding receptacles should be at least 12", but no more than 18", above ground level—requirements vary, so check with your local inspector.

2: Dig Trenches

When laying underground cables, save time and minimize lawn damage by digging trenches as narrow as possible. Plan the circuit to reduce the length of cable runs.

If your soil is sandy, or very hard and dry, water the ground thoroughly before you begin digging. Lawn sod can be removed, set on strips of plastic, and replaced after cables are laid. Keep the removed sod moist but not wet, and replace it within two or three days. Otherwise, the grass underneath the plastic may die.

If trenches must be left unattended, make sure to cover them with scrap pieces of plywood to prevent accidents and to keep water out.

Materials You Will Need:

Stakes, string, plastic, scrap piece of conduit, compression fittings, plastic bushings.

How to Dig Trenches for Underground Cables

1 Mark the outline of trenches with wooden stakes and string.

2 Cut two 18"-wide strips of plastic, and place one strip on each side of the trench outline.

3 Remove blocks of sod from the trench outline, using a shovel. Cut sod 2" to 3" deep to keep roots intact. Place the sod on one of the plastic strips, and keep it moist.

4 Dig the trenches to the depth required by your local Code. Heap the dirt onto the second strip of plastic.

5 To run cable under a sidewalk, cut a length of metal conduit about 1 ft. longer than width of sidewalk, then flatten one end of the conduit to form a sharp tip.

6 Drive the conduit through the soil under the sidewalk, using a ball peen or masonry hammer and a wood block to prevent damage to the pipe.

7 Cut off the ends of the conduit with a hacksaw, leaving about 2" of exposed conduit on each side. Underground cable will run through the conduit.

8 Attach a compression fitting and plastic bushing to each end of the conduit. The plastic fittings will prevent the sharp edges of the conduit from damaging the cable sheathing.

9 If trenches must be left unattended, temporarily cover them with scrap plywood to prevent accidents.

3: Install Boxes & Conduit

Use cast-aluminum electrical boxes for outdoor fixtures and install metal conduit to protect any exposed cables, unless your Code has different requirements. Standard metal and plastic electrical boxes are not watertight, and should never be used outdoors. A few local Codes require you to install conduit to protect all underground cables, but in most regions this is not necessary. Some local Codes allow you to use boxes and conduit made with PVC plastic.

Begin work by installing the retrofit boxes and the cables that run between them inside finished walls. Then install the outdoor boxes and conduit.

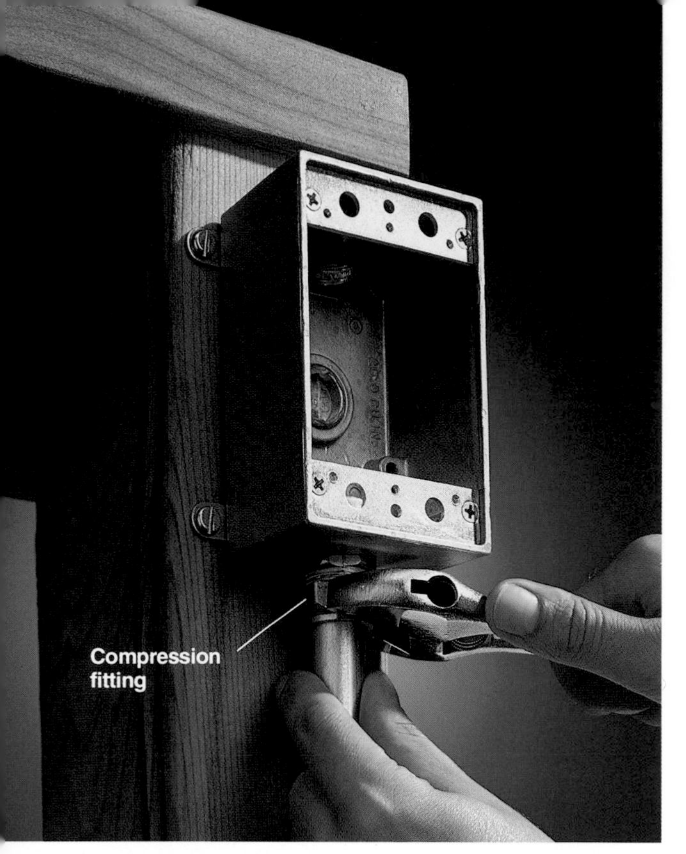

Electrical boxes for an outdoor circuit must be weatherproof. This outdoor receptacle box made of cast aluminum has sealed seams, and is attached to conduit with threaded, watertight compression fittings.

Materials You Will Need:

NM two-wire cable, cable staples, plastic retrofit light fixture box with grounding clip, plastic single-gang retrofit boxes with internal clamps, extension ring, silicone caulk, IMC or rigid metal conduit, pipe straps, conduit sweep, compression fittings, plastic bushings, Tapcon® anchors, single-gang outdoor boxes, galvanized screws, grounding pigtails, wire nuts.

How to Install Electrical Boxes & Conduit

1 Outline the GFCI receptacle box on the exterior wall. First drill pilot holes at the corners of the box outline, and use a piece of stiff wire to probe the wall for electrical wires or plumbing pipes. Complete the cutout with a jig saw or reciprocating saw.

Masonry variation: To make cutouts in masonry, drill a line of holes inside the box outline, using a masonry bit, then remove waste material with a masonry chisel and ball peen hammer.

2 From inside house, make the cutout for the indoor switch in the same stud cavity that contains the GFCI cutout. Outline the box on the wall, then drill a pilot hole and complete the cutout with a wallboard saw or jig saw.

3 On outside of house, make the cutout for the motion-sensor light fixture in the same stud cavity with the GFCI cutout. Outline the light fixture box on the wall, then drill a pilot hole and complete the cutout with a wallboard saw or jig saw.

4 Estimate the distance between the indoor switch box and the outdoor motion-sensor box, and cut a length of NM cable about 2 ft. longer than this distance. Use a fish tape to pull the cable from the switch box to the motion-sensor box.

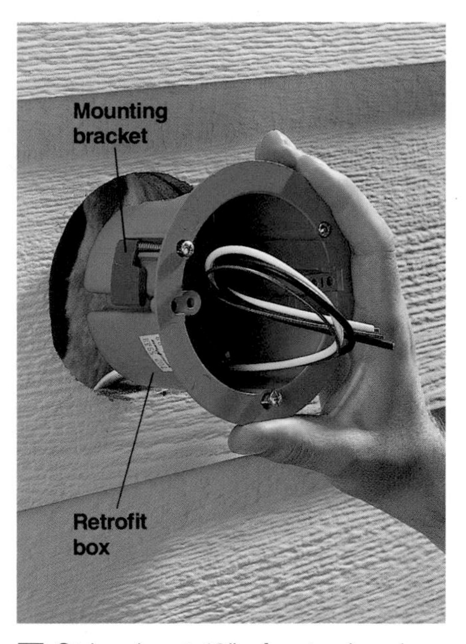

5 Strip about 10" of outer insulation from the end of the cable, using a cable ripper. Open a knockout in the retrofit light fixture box with a screwdriver. Insert the cable into the box so that at least 1/4" of outer sheathing reaches into the box.

6 Insert the box into the cutout opening, and tighten the mounting screws until the brackets draw the outside flange firmly against the siding.

7 Estimate the distance between the outdoor GFCI cutout and the indoor switch cutout, and cut a length of NM cable about 2 ft. longer than this distance. Use a fish tape to pull the cable from the GFCI cutout to the switch cutout. Strip 10" of outer insulation from both ends of each cable.

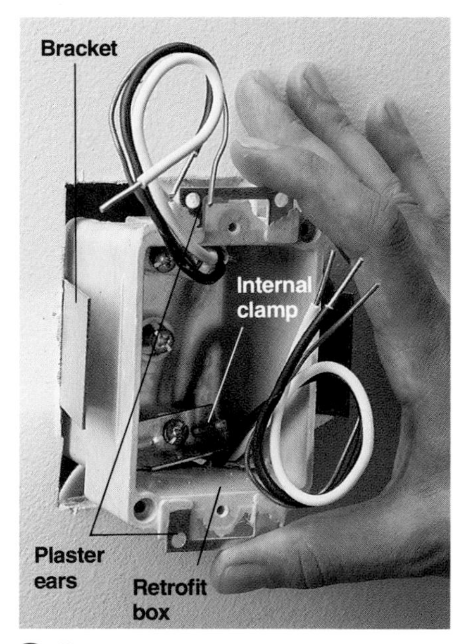

8 Open one knockout for each cable that will enter the box. Insert the cables so at least 1/4" of outer sheathing reaches inside box. Insert box into cutout, and tighten the mounting screw in the rear of the box until the bracket draws the plaster ears against the wall. Tighten internal cable clamps.

(continued next page)

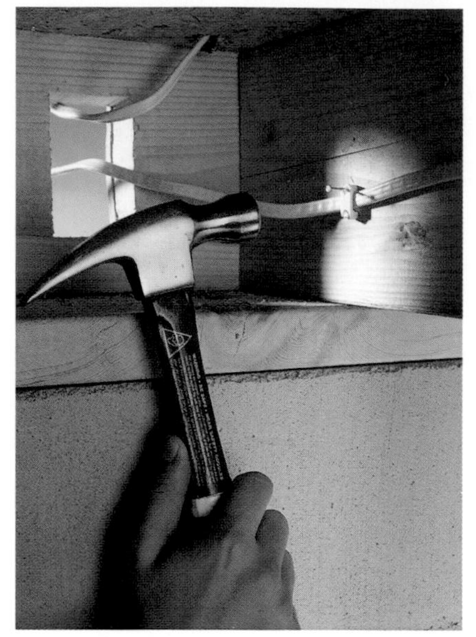

9 Install NM cable from circuit breaker panel to GFCI cutout. Allow an extra 2 ft. of cable at panel end, and an extra 1 ft. at GFCI end. Attach cable to framing members with cable staples. Strip 10" of outer sheathing from the GFCI end of cable, and 3/4" of insulation from each wire.

10 Open one knockout for each cable that will enter the GFCI box. Insert the cables so at least 1/4" of sheathing reaches into the box. Push the box into the cutout, and tighten the mounting screw until the bracket draws the plaster ears tight against the wall.

11 Position a foam gasket over the GFCI box, then attach a extension ring to the box, using the mounting screws included with the extension ring. Seal any gaps around the extension ring with silicone caulk.

12 Measure and cut a length of IMC conduit to reach from the bottom of the extension ring to a point about 4" from the bottom of the trench. Attach the conduit to the extension ring using a compression fitting.

13 Anchor the conduit to the wall with a pipe strap and Tapcon® screws. Or, use masonry anchors and pan-head screws. Drill pilot holes for anchors, using a masonry drill bit.

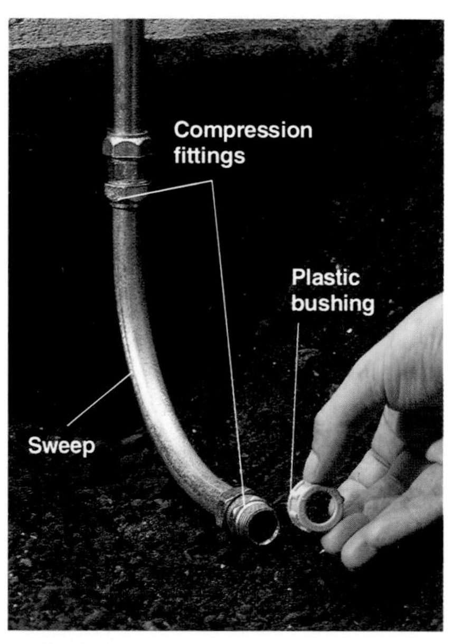

14 Attach compression fittings to the ends of metal sweep fitting, then attach the sweep fitting to the end of the conduit. Screw a plastic bushing onto the exposed fitting end of the sweep to keep the metal edges from damaging the cable.

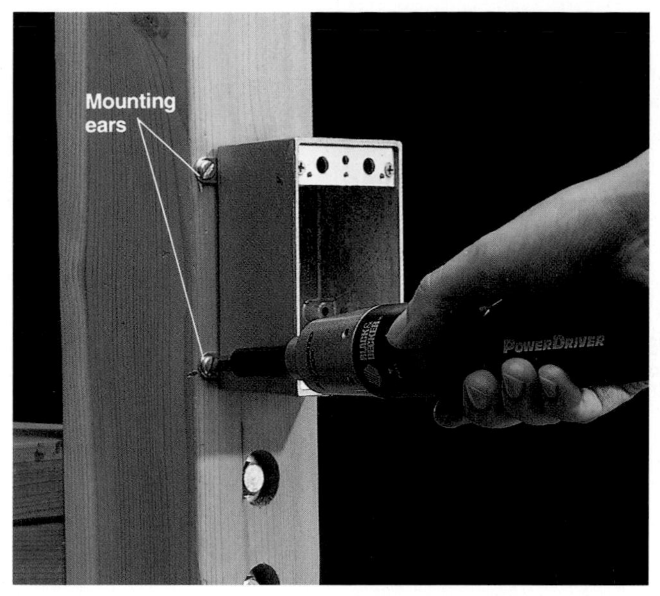

15 Attach mounting ears to the back of a weatherproof receptacle box, then attach the box to the deck frame by driving galvanized screws through the ears and into the post.

16 Measure and cut a length of IMC conduit to reach from the bottom of the receptacle box to a point about 4" from the bottom of the trench. Attach the conduit to the box with a compression fitting. Attach a sweep fitting and plastic bushing to the bottom of the conduit, using compression fittings (see step 14).

17 Cut a length of IMC conduit to reach from the top of the receptacle box to the switch box location. Attach the conduit to the receptacle box with a compression fitting. Anchor the conduit to the deck frame with pipe straps.

18 Attach mounting ears to the back of switch box, then loosely attach the box to the conduit with a compression fitting. Anchor the box to the deck frame by driving galvanized screws through the ears and into the wood. Then tighten the compression fitting with a wrench.

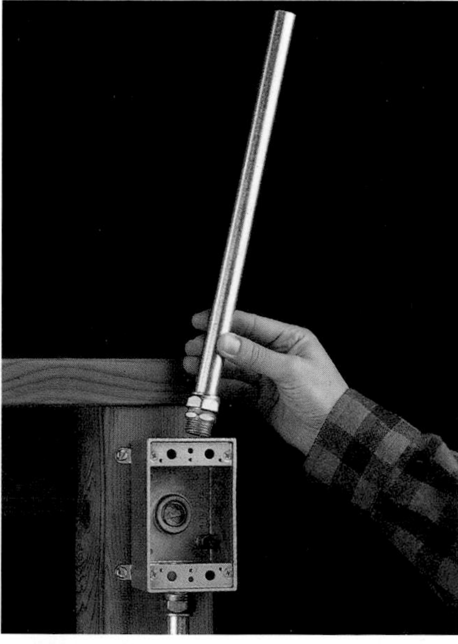

19 Measure and cut a short length of IMC conduit to reach from the top of the switch box to the deck light location. Attach the conduit with a compression fitting.

4: Install UF Cable

Use UF cable for outdoor wiring if the cable will come in direct contact with soil. UF cable has a solid-core vinyl sheathing, and cannot be stripped with a cable ripper. Instead, use a utility knife and the method shown (steps 5 & 6, page opposite). Never use NM cable for outdoor wiring. If your local Code requires that underground wires be protected by conduit, use THHN/THHW wire instead of UF cable.

After installing all cables, you are ready for the rough-in inspection. While waiting for the inspector, temporarily attach the weatherproof coverplates to the boxes, or cover them with plastic to prevent moisture from entering. After the inspector has approved the rough-in work, fill in all cable trenches and replace the sod before making the final connections.

Materials You Will Need:

UF cable, electrical tape, grounding pigtails, wire nuts, weatherproof coverplates.

How to Install Outdoor Cable

1 Measure and cut all UF cables, allowing an extra 12" at each box. At each end of the cable, use a utility knife to pare away about 3" of outer sheathing, leaving the inner wires exposed.

2 Feed a fish tape down through the conduit from the GFCI box. Hook the wires at one end of the cable through the loop in the fish tape, then wrap electrical tape around the wires up to the sheathing. Carefully pull the cable through the conduit.

3 Lay the cable along the bottom of the trench, making sure it is not twisted. Where cable runs under a sidewalk, use the fish tape to pull it through the conduit.

4 Use the fish tape to pull the end of the cable up through the conduit to the deck receptacle box at the opposite end of the trench. Remove the cable from the fish tape.

5 Cut away the electrical tape at each end of the cable, then clip away the bent wires. Bend back one of the wires in the cable, and grip it with needlenose pliers. Grip the cable with another pliers.

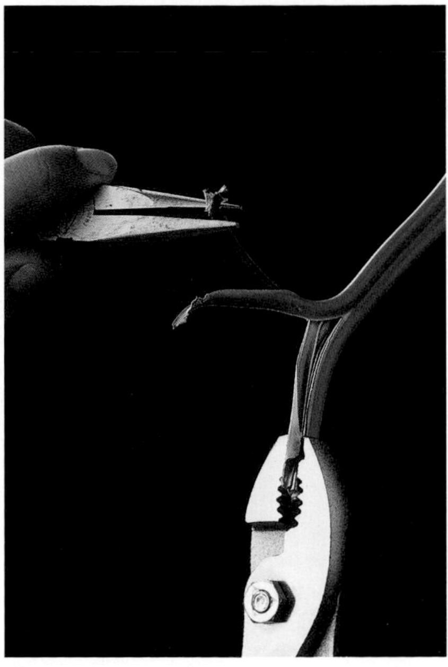

6 Pull back on the wire, splitting the sheathing and exposing about 10" of wire. Repeat with the remaining wires, then cut off excess sheathing with a utility knife. Strip 3/4" of insulation from the end of each wire, using a combination tool.

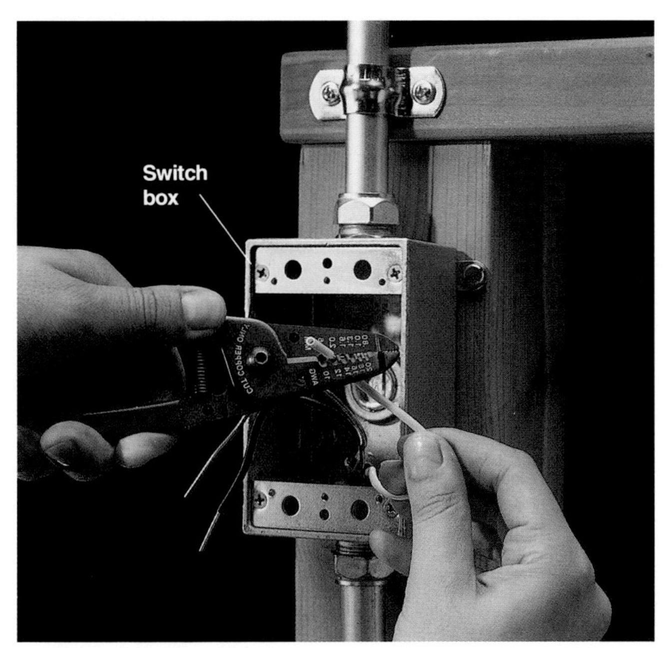

7 Measure, cut, and install a cable from the deck receptacle box to the outdoor switch box, using the fish tape. Strip 10" of sheathing from each end of the cable, then strip 3/4" of insulation from the end of each wire, using a combination tool.

8 Attach a grounding pigtail to the back of each metal box and extension ring. Join all grounding wires with a wire nut. Tuck the wires inside the boxes, and temporarily attach the weatherproof coverplates until the inspector arrives for the rough-in inspection.

Arrange for the rough-in inspection before making the final connections.

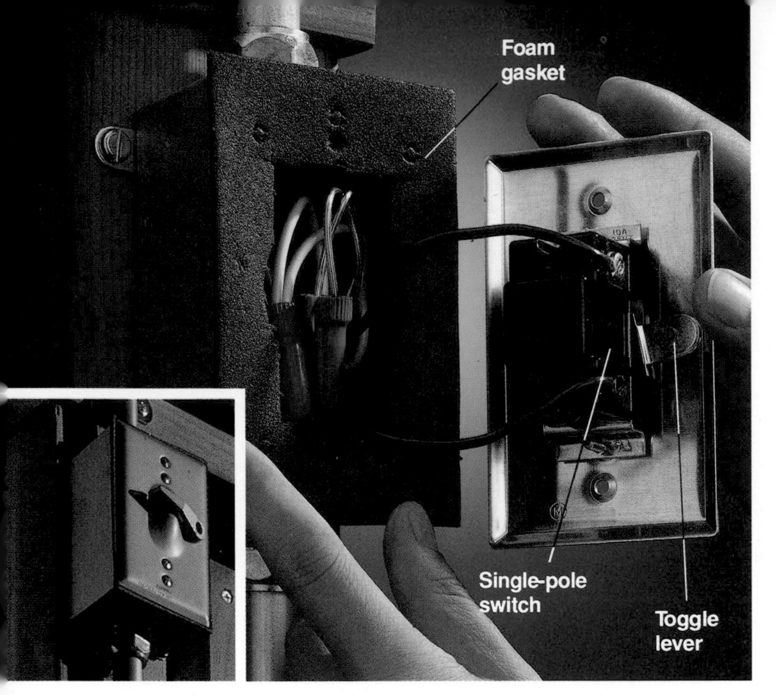

Switches for outdoor use have weatherproof coverplates with built-in toggle levers. The lever operates a single-pole switch mounted to the inside of the coverplate. Connect the black circuit wire to one of the screw terminals on the switch, and connect the black wire lead from the light fixture to the other screw terminal. Use wire nuts to join the white circuit wires and the grounding wires.

5: Make Final Connections

Make the final hookups for the switches, receptacles, and light fixtures after the rough-in cable installation has been reviewed and approved by your inspector, and after all trenches have been filled in. Install all the light fixtures, switches, and receptacles, then connect the circuit to the circuit breaker panel.

Because outdoor wiring poses a greater shock hazard than indoor wiring, the GFCI receptacle (page 70) in this project is wired to provide shock protection for all fixtures controlled by the circuit.

When all work is completed and the outdoor circuit is connected at the service panel, your job is ready for final review by the inspector.

Materials You Will Need:

Motion-sensor light fixture, GFCI receptacle, 15-amp grounded receptacle, outdoor switch, decorative light fixture, wire nuts.

How to Connect a Motion-sensor Light Fixture

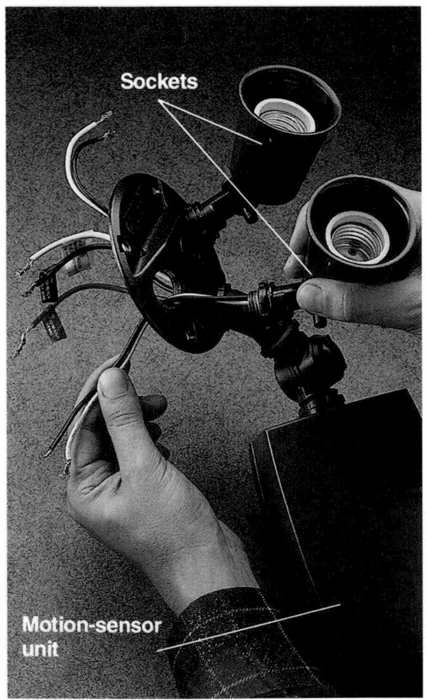

1 Assemble fixture by threading the wire leads from the motion-sensor unit and the bulb sockets through the faceplate knockouts. Screw the motion-sensor unit and bulb sockets into the faceplate.

2 Secure the motion-sensor unit and the bulb sockets by tightening the locknuts.

3 Insert the fiber washers into the sockets, and fit a rubber gasket over the end of each socket. The washers and gaskets ensure that the fixture will be watertight.

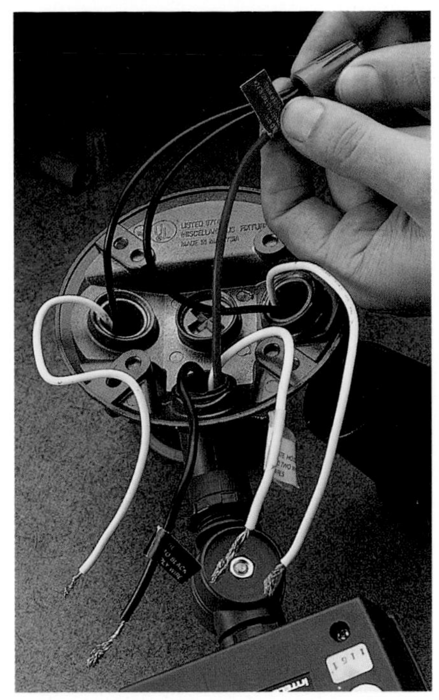

4 Connect the red wire lead from the motion-sensor unit to the black wire leads from the bulb sockets, using a wire nut. Some light fixtures have pre-tagged wire leads for easy installation.

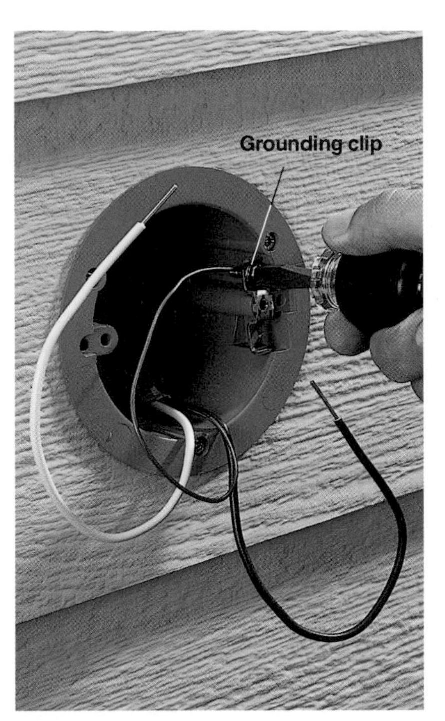

5 Attach the bare copper grounding wire to the grounding clip on the box.

6 Slide the foam gasket over the circuit wires at the electrical box. Connect the white circuit wire to the white wire leads on the light fixture, using a wire nut.

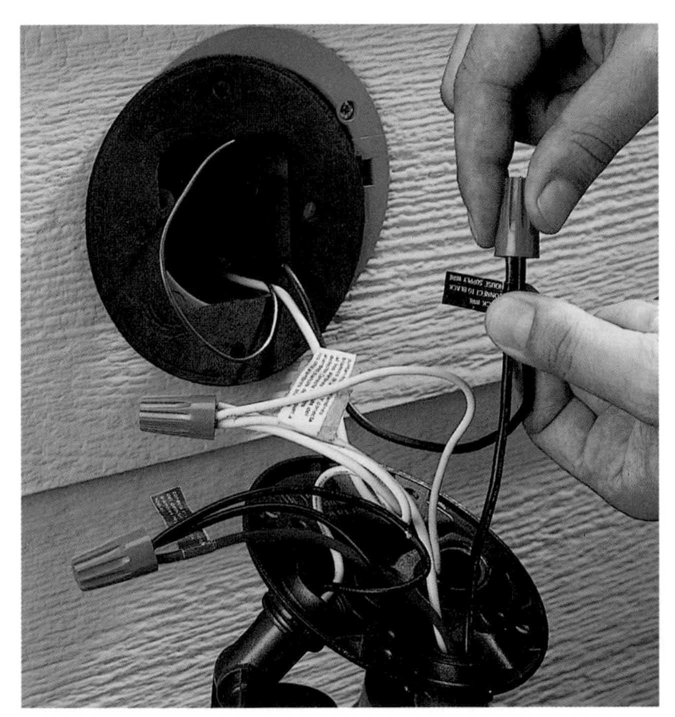

7 Connect the black circuit wire to the black wire lead on the light fixture, using a wire nut.

8 Carefully tuck the wires into the box, then position the light fixture and attach the faceplate to the box, using the mounting screws included with the light fixture.

How to Connect the GFCI Receptacle

1 Connect the black feed wire from the power source to the brass terminal marked LINE. Connect the white feed wire from the power source to the silver screw terminal marked LINE.

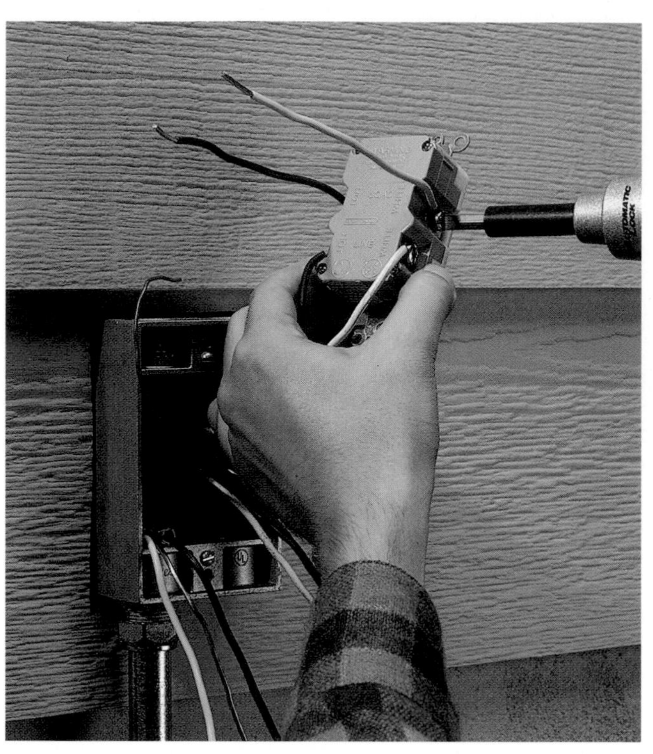

2 Attach a short white pigtail wire to the silver screw terminal marked LOAD, and attach a short black pigtail wire to the brass screw terminal marked LOAD.

3 Connect the black pigtail wire to all the remaining black circuit wires, using a wire nut. Connect the white pigtail wire to the remaining white circuit wires.

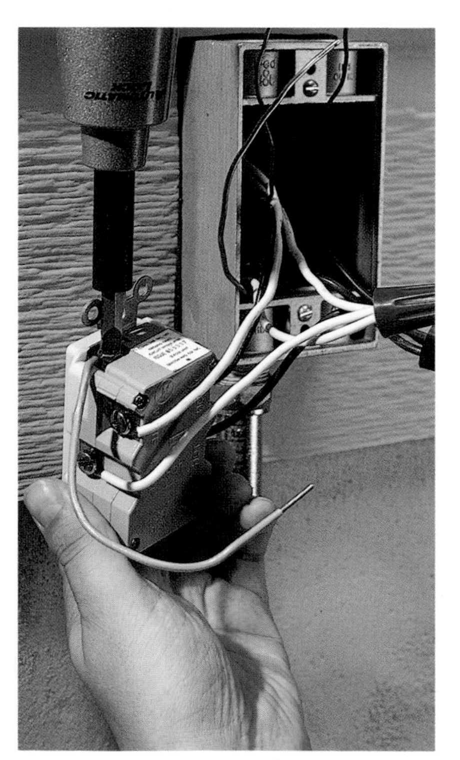

4 Attach a grounding pigtail to the grounding screw on the GFCI. Join the grounding pigtail to the bare copper grounding wires, using a wire nut.

5 Carefully tuck wires into box. Mount GFCI, then fit a foam gasket over the box and attach the weatherproof coverplate.

How to Connect an Outdoor Receptacle

1 Connect the black circuit wires to the brass screw terminals on the receptacle. Connect the white circuit wires to the silver screw terminals on the receptacle. Attach a grounding pigtail to the grounding screw on the receptacle, then join all grounding wires with a wire nut.

2 Carefully tuck all wires into the box, and attach the receptacle to the box, using the mounting screws. Fit a foam gasket over the box, and attach the weatherproof coverplate.

How to Connect a Decorative Light Fixture

1 Thread the wire leads of the light fixture through a threaded compression fitting. Screw the union onto the base of the light fixture.

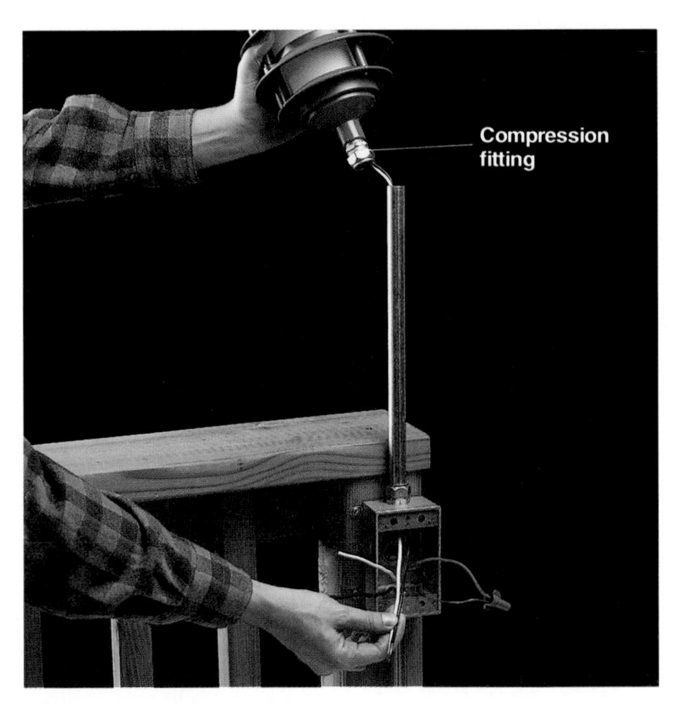

2 Feed wire leads through conduit and into switch box. Slide light fixture onto conduit, and tighten compression fitting. Connect black wire lead to one screw terminal on switch, and connect white wire lead to white circuit wire.

Make hookups at circuit breaker panel and arrange for final inspection.

KOHLER

Advanced Plumbing Projects

Upgrading your home's plumbing system offers substantial rewards. Not only does new plumbing make your home more livable, it can also greatly increase its market value. Installing new plumbing is a perfect do-it-yourself project, because the work, though time-consuming, is not particularly difficult. You also can save hundreds or even thousands of dollars by doing the work, rather than using plumbing contractors.

This section of *Advanced Wiring & Plumbing Projects* is designed to meet the needs of an intermediate to advanced-level do-it-yourselfer with some experience working with plumbing materials. If you've made your own minor plumbing repairs—replaced a faucet, fixed a toilet ballcock, patched a leaking pipe—then the projects in this book are well within your capabilities. (Be aware that many plumbing projects also require basic carpentry techniques.) You will learn everything you need to know to install new plumbing lines, whether in a brand-new room addition or a converted space undergoing major remodeling. You also will learn techniques for replacing old steel and cast-iron pipes with new copper and plastic pipes.

First you will learn the basic mechanics of the plumbing network in your house, and how to inspect and map your existing plumbing system. You will see how to select and work with a variety of plastic and copper pipes and fittings required for advanced plumbing projects. You also will learn about Plumbing Codes and how to work with building officials while planning your project.

Next, *Installing New Plumbing* uses four carefully chosen demonstration projects to show you how to install new plumbing in various situations, including a remodeled bathroom or kitchen, and a utility sink in a detached garage. No two projects are alike, so it is impossible to cover every possible situation. When read as a group, however, these demonstrations show all the principles and techniques you are likely to need when completing your own plumbing project.

Replacing Old Plumbing demonstrates the techniques for running new plumbing pipes when walls and floors are finished. You will learn how to replace all plumbing pipes between the water meter and the individual fixtures. Some demolition and repair work is inevitable in any major plumbing project, but you will see methods used by professional plumbers to save time and simplify this task.

This section of *Advanced Wiring & Plumbing Projects* takes the mystery out of understanding and installing a plumbing system. It will help you create a home environment that exactly meets your needs, and help you save money doing so.

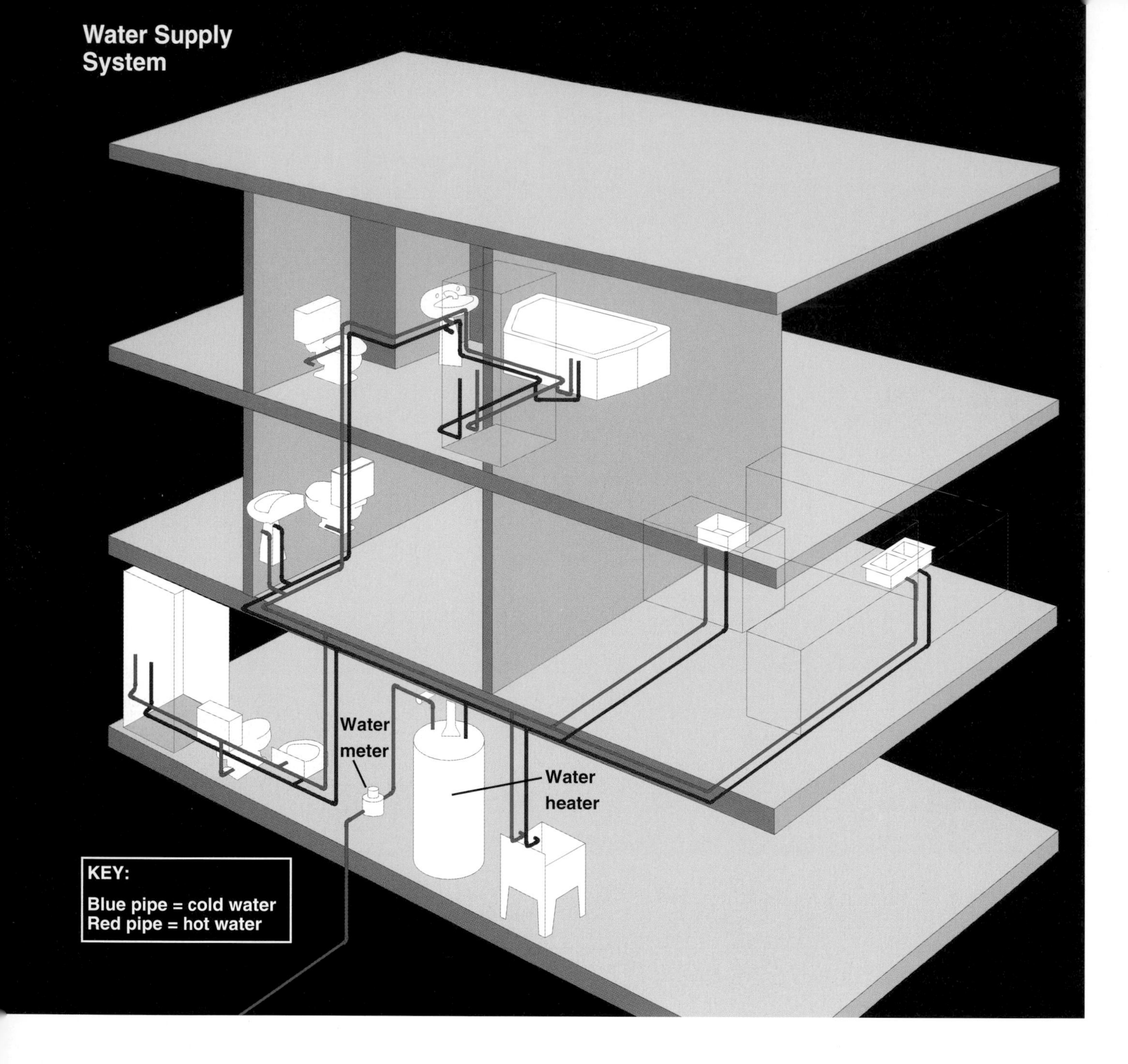

Using Water in Your Home

Behind the walls, under the floors, and beneath the landscape, a typical home hides a seemingly chaotic system of pipes, drains, hoses, tubes, sprinklers, valves, elbows, traps, bends, valves, faucets, fixtures, and spigots—enough to nearly circle a football field, if all the pipes and fittings were laid end to end.

But no matter how bewildering the plumbing network appears, its purpose is simple: to carry fresh water to points of use and to transport waste water out of your home. All home plumbing systems consist of two separate systems: the water supply system (above) and the drain-waste-vent (DWV) system (opposite page).

Water supply pipes are relatively small—1" or less in diameter. They form a tightly sealed, pressurized system controlled by valves and faucets. Older homes may have galvanized steel supply pipes, while newer homes (or renovated older homes) have copper or plastic supply pipes.

Upon arriving in your home, the 1" main water supply pipe passes through the water meter and begins to split into ¾" or ½" branch lines almost immediately. One branch passes into a water heater, which will supply hot water for faucets and other fixtures in the home. Hot and cold water branch pipes run parallel to one another on their way to kitchens, bathrooms, and utility sinks

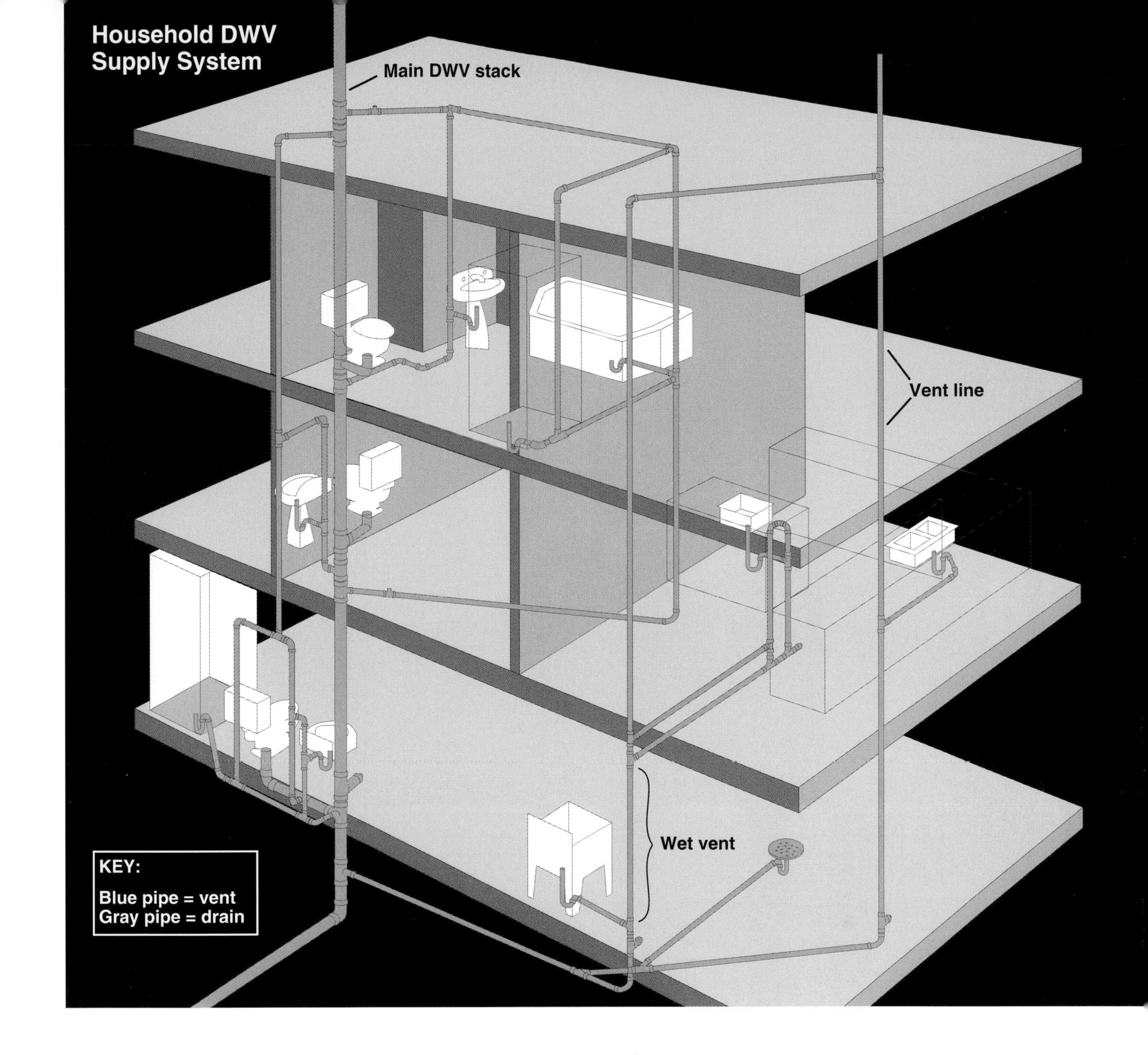

throughout the home. Other branch lines, supplying only cold water, extend through the foundation to outdoor hose bibs and detached outbuildings.

DWV pipes, unlike the sealed supply system, form an open system ventilated to the outdoors. At 1¼" to 4" in diameter, DWV pipes are larger than the water supply lines. In older homes, DWV pipes are typically cast iron or galvanized steel, while in newer houses they are usually plastic.

To prevent noxious gases from rising into your home from the sewer, each fixture in your DWV system is serviced by a looped section of pipe, called a *drain trap*. The trap holds standing water, effectively sealing the sewer system off from the interior of your home. If the system were sealed, flowing waste water would create a partial vacuum that could suck the standing water out of drain traps. To keep the system open and prevent this problem, each fixture drain is connected to a nearby vent line, which is linked to a waste-vent stack extending through the roof. By equalizing pressure in the system, the vent pipes ensure the proper function and safety of your drain system.

Mapping Your Plumbing System

Mapping your home's plumbing system is a good way to familiarize yourself with the plumbing layout and can help you when planning plumbing renovation projects. With a good map, you can envision the best spots for new fixtures and plan new pipe routes more efficiently. Maps also help in emergencies, when you need to locate burst or leaking pipes quickly.

Draw a plumbing map for each floor on tracing paper, so you can overlay floors and still read the information below. Make your drawings to scale and have all plumbing fixtures marked. Fixture templates and tracing paper are available at drafting supply stores.

Find supply lines inside walls or beneath floors by listening for running water with a stethoscope or a drinking glass while a helper runs water from the fixture. Use fixture locations on the floors above and below to find the general location of pipes.

Tips for Mapping Your Plumbing System

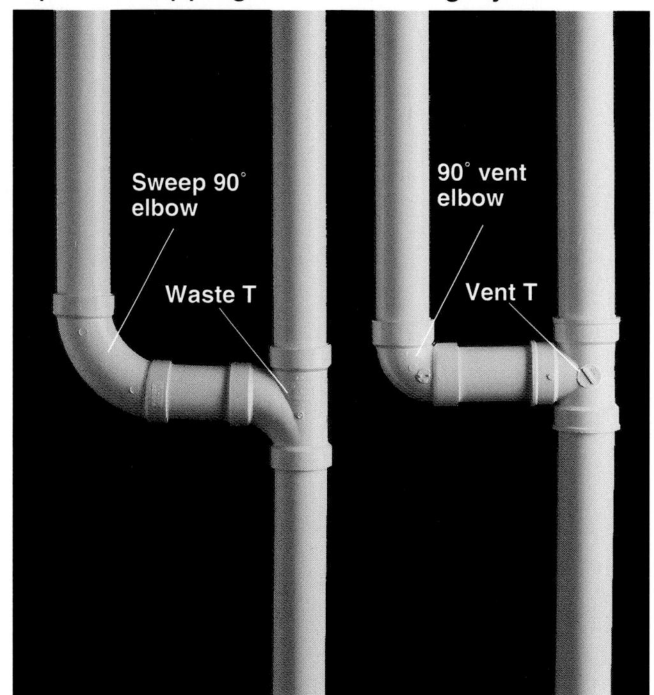

Identify drains and vents by the shape of their fittings. Drain pipes (left) require gradual changes in direction, requiring the use of Y-fittings, waste T-fittings, and sweep 90° elbows. Vents (right) can use fittings with abrupt changes in direction, such as vent T-fittings and vent elbows.

Pinpoint the location of the main stack when all interior walls are finished by jiggling a hand auger down the roof vent while a helper listens to walls from inside the house. Always exercise caution when working on a roof.

Use floor plans of your house to create your plumbing map. Convert the general outlines for each story to tracing paper. The walls can be drawn larger than scale to fit all the plumbing symbols you will map, but keep overall room dimensions and plumbing fixtures to scale. Be sure to make diagrams for basements and attic spaces as well.

Locate and map all valves throughout the supply lines. This will allow you to shut off only the necessary branches when making repairs, while maintaining service to the rest of the house. Use the correct symbols (right) to identify different valve types (page 83).

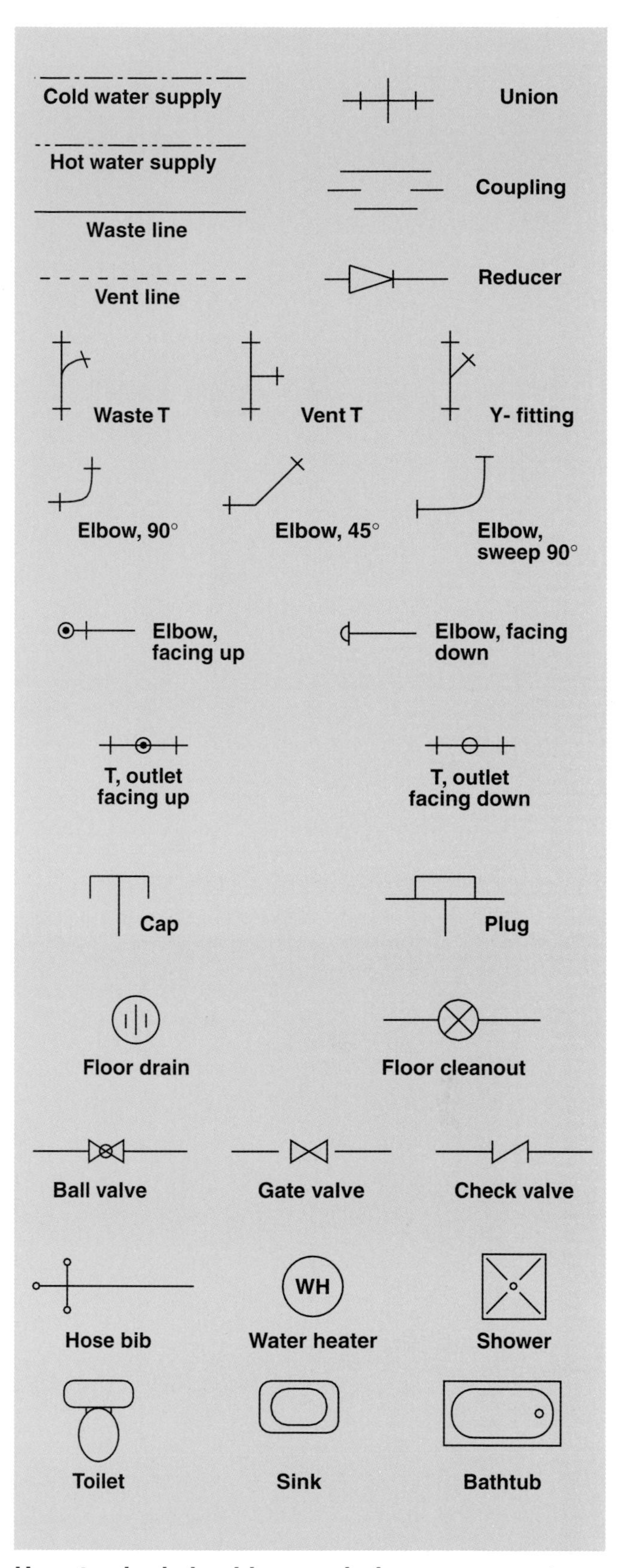

Use standard plumbing symbols on your map to identify the components of your plumbing system. These symbols will help you and your building inspector follow connections and transitions more easily.

How to Map Water Supply Pipes

1 Locate the water meter, usually found along a basement wall. The meter is the first main fitting in the supply line. Mark its location on your basement diagram.

2 Follow the cold water distribution pipe past the main shutoff valve to the water heater, generally the first appliance to receive water. Map the valve and water heater locations on the basement diagram.

3 Locate cold water branch pipes leading to sillcocks, which supply water to hose bibs outside the house. Indicate these branches on the basement map.

4 Return to the water heater and map the location of hot and cold supply lines running to basement utility fixtures, such as a washing machine and utility sink.

5 Map the routes to any remaining basement plumbing fixtures on your basement diagram. Pipe runs that serve both basement and first-floor plumbing should be marked on both the basement and first-story maps.

6 Find where vertical supply pipe risers extend up into the floors above. Supply lines generally rise straight to the first-floor fixtures. If there is doubt, measure from the nearest outside wall to the supply line riser, and do the same at the respective first-floor fixture. If measurements are not the same, there is a hidden offset in the pipe route.

7 Map the supply routes to all first-story fixtures by laying the first-floor diagram over the basement map and transferring the locations of vertical supply risers. Indicate any jogs in the supply lines occurring between floors.

8 Overlay the second-story diagram over the first-floor map, and mark the location of supply pipes—generally they will extend directly up from fixtures below. If first-story and second-story fixtures are not closely aligned, the supply pipes follow an offset route in wall or floor cavities. By overlaying the maps, you can see the relation and distance between fixtures and accurately estimate pipe routes. If no obvious path exists for supply pipes, try locating the pipes by listening in likely areas with a stethoscope (page 76).

How to Map DWV Pipes

1 From the basement, locate the main waste-vent stack and any fixtures that drain directly into it, such as a basement toilet.

2 Determine the path of the main drain under the basement slab by following the main stack to the cleanout hub on the basement floor. The cleanout is usually located near a basement wall facing the street.

3 Note any auxiliary waste-vent stacks that enter the basement floor. These are typically 2"-diameter pipes, compared to 3"- or 4"-diameter main waste-vent stacks. Auxiliary waste-vent stacks are often located near basement utility sinks or below a kitchen located far from the main stack.

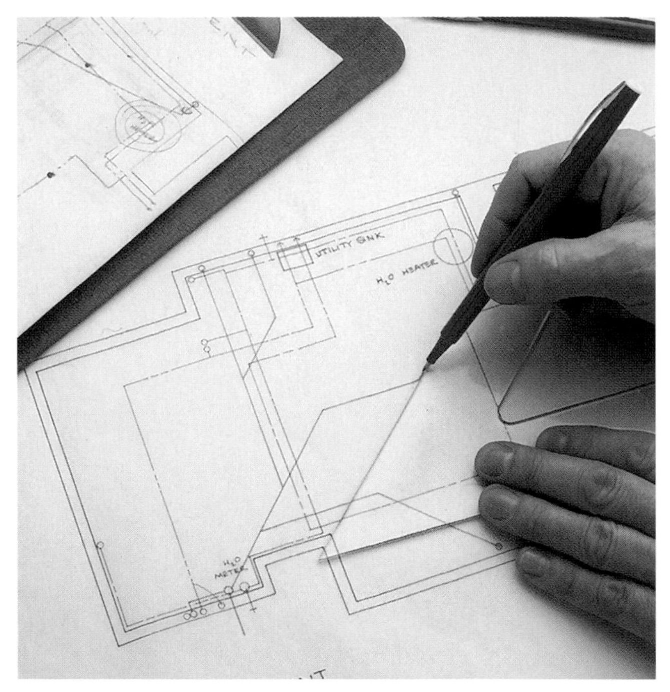

4 On your basement diagram, map the location of the main waste-vent stack, the cleanout hub, and the horizontal main drain pipe. Also note the location of auxiliary stacks, and estimate the path of the horizontal drain pipe connecting the auxiliary stacks to the main drain.

5 From the basement, note the location of horizontal drain pipes running overhead, and the points where vertical drain pipes extend up into the floors above. Overlay your first-story diagram onto the basement map, then transfer the location of the vertical waste-vent stacks. Mark the location of all horizontal drain pipes running below the floor.

6 Overlay the second-story diagram over the basement map, transfer the location of the vertical waste-vent stacks, and mark the location of any horizontal drain pipes running beneath the floor. Since the floor spaces between the first and second story are usually finished, you may need to estimate their location. These horizontal drain pipes will usually drain into the nearest waste-vent stack.

7 Finally, map the location of vent pipes as accurately as you can. If possible, look in your attic to determine where vent pipes emerge from the story below. Indicate whether the individual vent pipes connect to a waste-vent stack or extend through the roofline.

Plumbing Materials

The materials used in home plumbing systems are closely regulated by Building Codes. The materials shown here and on the following pages are approved for use by the National Uniform Plumbing Code at the time this book was published. However it is possible that your local code has other requirements, so always check with your local building officials. Approved materials are stamped with one or more product standard codes. Look for these stamps when buying your pipes and fittings.

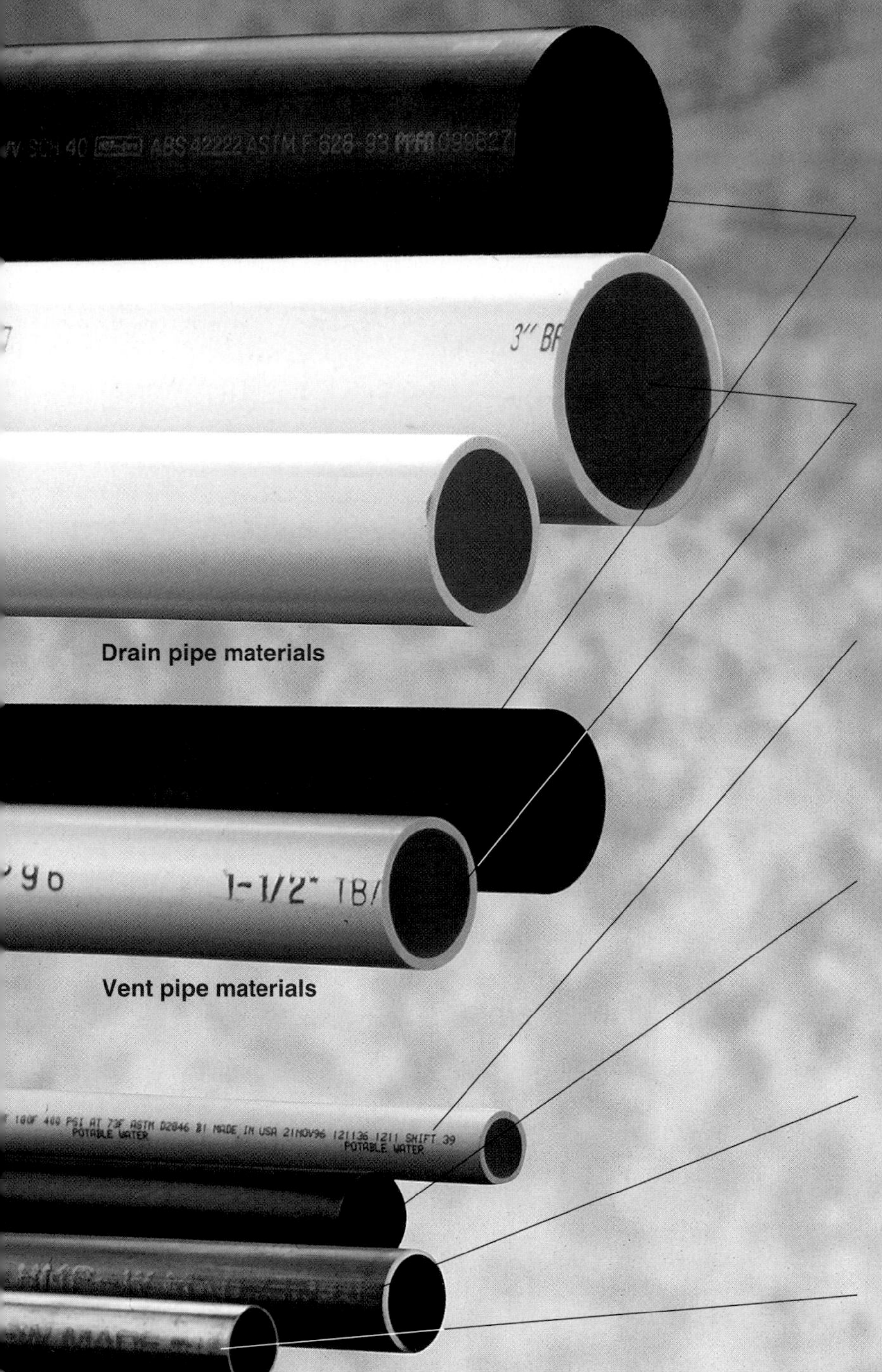

Drain pipe materials

Vent pipe materials

Supply pipe materials

Common Pipe Materials

Schedule 40 ABS (acrylonitrile butadiene styrene) plastic is a rigid black or dark gray pipe used for drain and vent lines. It is commonly available in 10-ft. and 20-ft. lengths, in diameters of 1½", 2", 3", and 4".

Schedule 40 PVC (polyvinyl chloride) plastic is a white or cream-colored rigid pipe most commonly used for drain and vent lines. PVC is sold in 10-ft. and 20-ft. lengths, in diameters of 1¼", 1½", 2", 3", and 4".

CPVC (chlorinated polyvinyl chloride) plastic is a cream-colored pipe that in some areas is approved for use in hot and cold water supply lines. Where allowed, CPVC should be rated for 150 psi and 210°F when used in water supply lines. It is commonly sold in 10- ft. and 20-ft. lengths, in diameters of ½", ¾", and 1".

PE (polyethylene) plastic is a black or bluish flexible pipe that is often used for main water service lines from the street to the home, and for outdoor cold water pipes, such as those used for landscape watering systems. PE pipe is sold in coils of 50 ft. or more, in ½", ¾", and 1" diameters.

Type-L copper is a thick-walled pipe used primarily for underground water supply lines. Type-L is available both in rigid and semi-flexible form. Rigid copper is sold in 10-ft. and 20-ft. lengths; flexible copper in 60-ft. coils. Type-L copper is available in ½", ¾" and 1" diameters.

Type-M copper is the standard for indoor water supply lines. It has thinner walls than type-L copper, making in less expensive and easier to cut. It is available in 10-ft. and 20-ft. lengths, in diameters of ½", ¾" and 1".

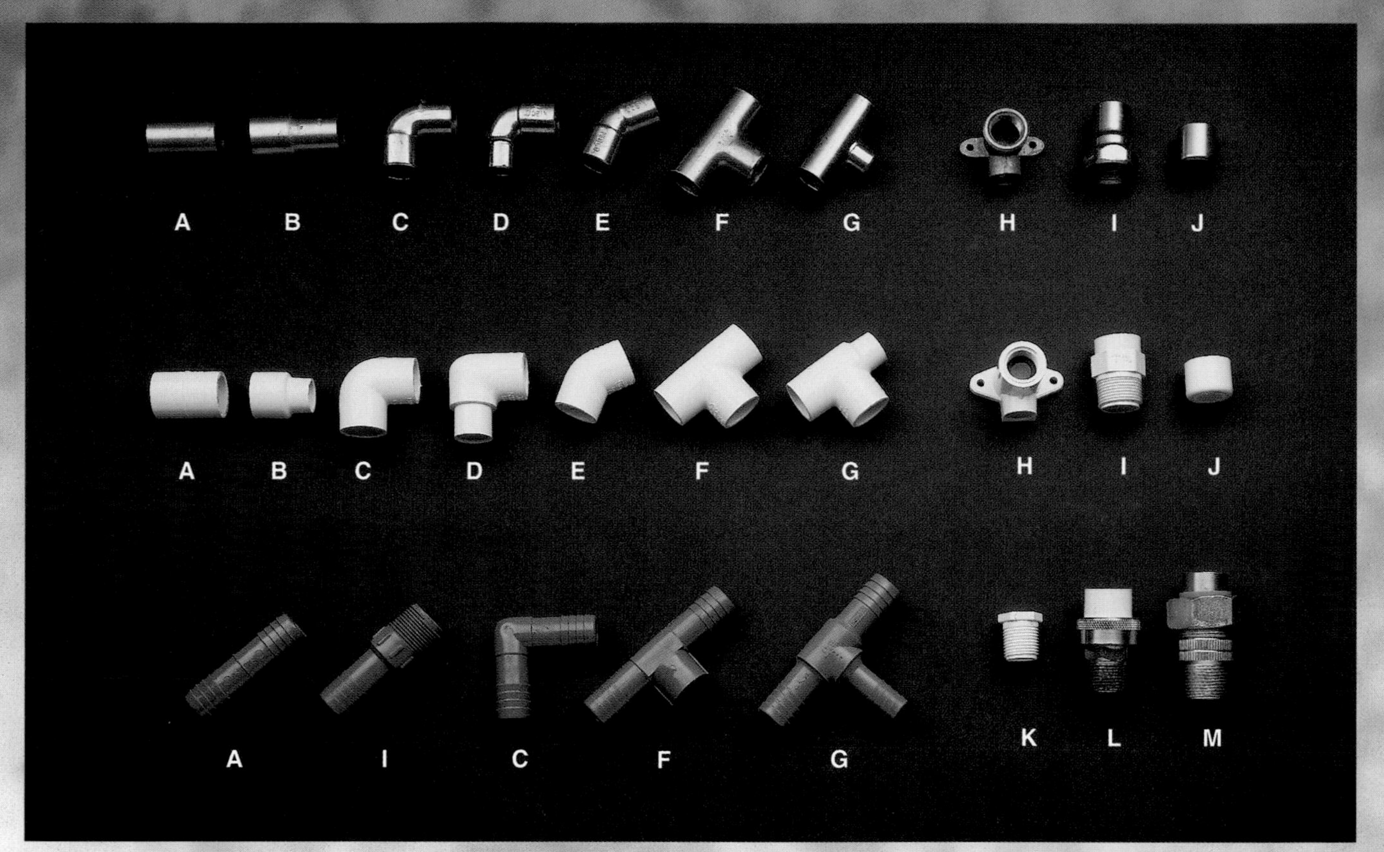

Water supply fittings are available in copper (top), CPVC plastic (center), and PVC plastic (bottom). PVC water supply fittings are gray with barbed sleeves, and are used only with cold water PE pipe. Fittings for each material are available in many shapes, including: unions (A), reducers (B), 90° elbows (C), reducing elbows (D), 45° elbows (E), T-fittings (F), reducing T-fittings (G), drop ear elbows (H), threaded adapters (I), caps (J), plug (K), CPVC to copper transition (L), and copper to steel transition (M).

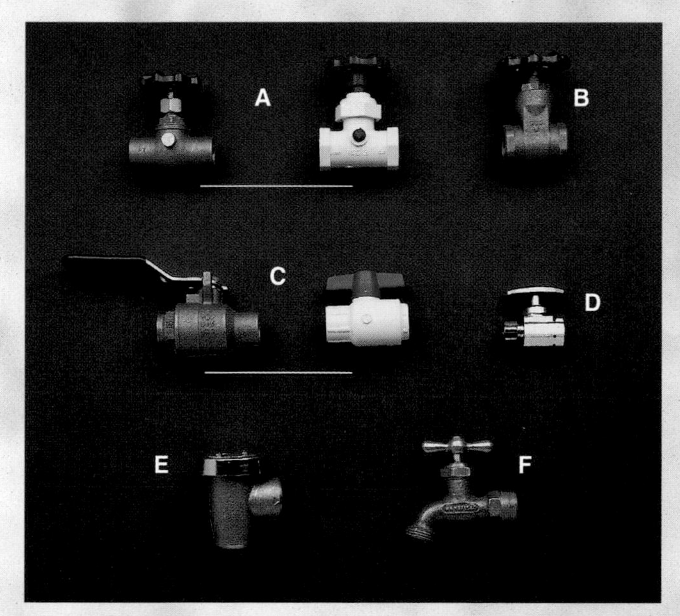

Water supply valves are available in bronze or plastic and in a variety of styles, including: drain-and-waste valves (A), gate valve (B), full-bore ball valves (C), fixture shutoff valve (D), vacuum breaker (E), and hose bibb (F).

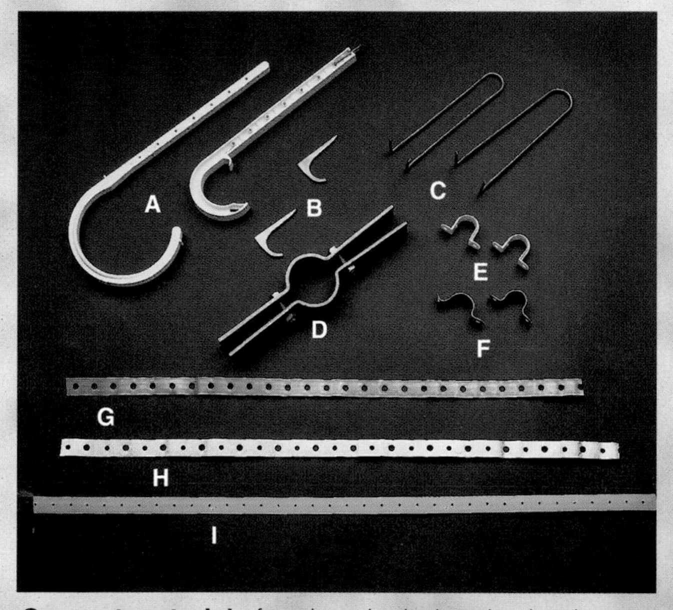

Support materials for pipes include: plastic pipe hangers (A), copper J-hooks (B), copper wire hangers (C), riser clamp (D), copper pipe straps (E), plastic pipe straps (F), flexible copper, steel, and plastic pipe strapping (G, H, I). Do not mix metal types when supporting metal pipes: use copper support materials for copper pipe, steel for steel and cast-iron pipes.

DWV Fittings

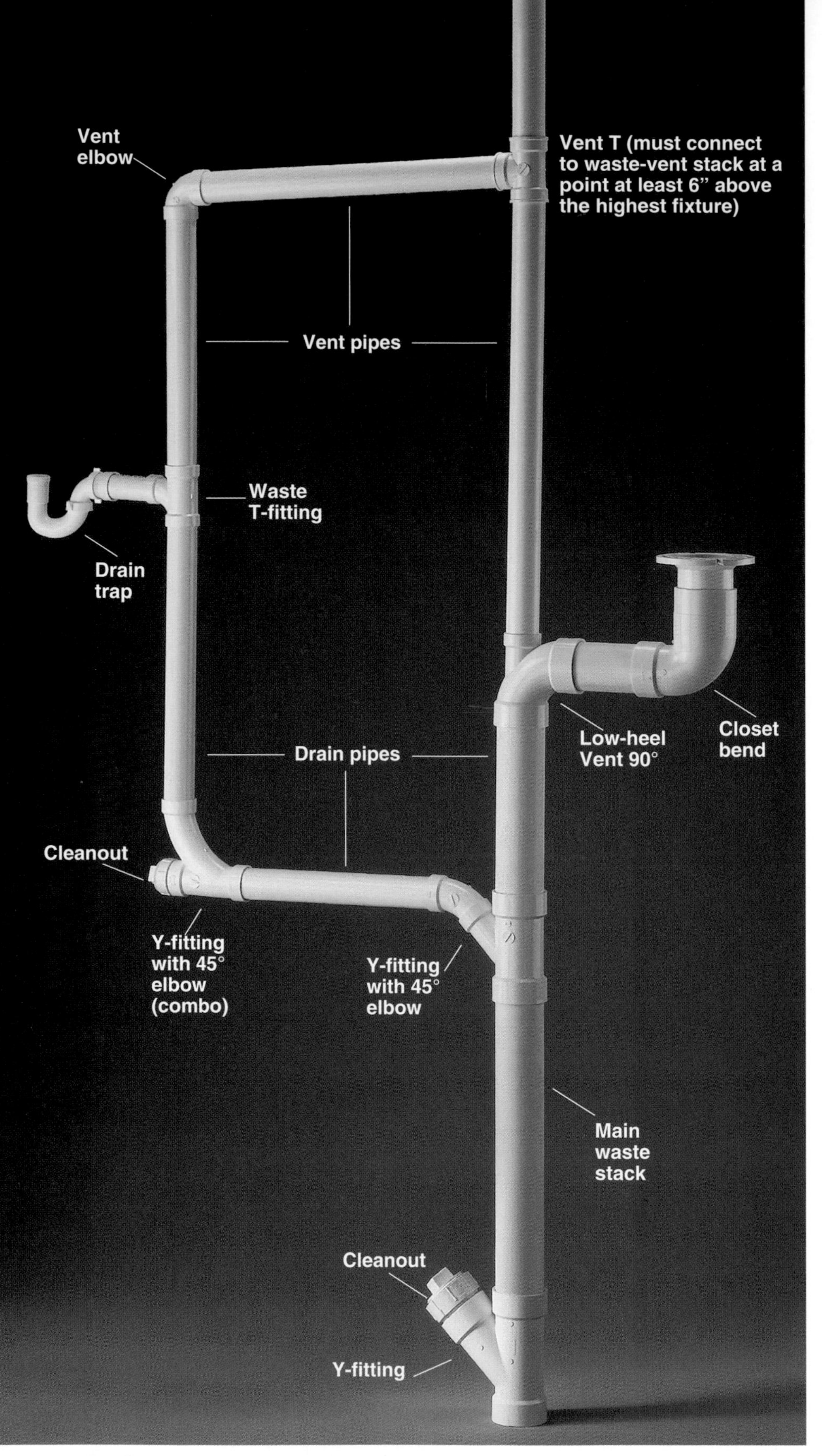

Basic DWV tree shows the correct orientation of drain and vent fittings in a plumbing system. Bends in vent pipes can be very sharp, but drain pipes should use fittings with a noticeable sweep. Fittings used to direct falling waste water from a vertical to a horizontal pipe should have bends that are even more sweeping. Your local Plumbing Code may require that you install cleanout fittings where vertical drain pipes meet horizontal runs.

Use the photos on these pages to identify the DWV fittings specified in the project how-to directions found later in this book. Each fitting shown is available in a variety of sizes to match your needs. Always use fittings made from the same material as your DWV pipes.

DWV fittings come in a variety of shapes to serve different functions within the plumbing system.

Vents: In general, the fittings used to connect vent pipes have very sharp bends with no sweep. Vent fittings include the vent T and vent 90° elbow. Standard drain pipe fittings can also be used to join vent pipes.

Horizontal-to-vertical drains: To change directions in a drain pipe from the horizontal to the vertical, use fittings with a noticeable sweep. Standard fittings for this use include waste T-fittings and 90° elbows. Y-fittings and 45° and 22° elbows can also be used for this purpose.

Vertical-to-horizontal drains: To change directions from the vertical to the horizontal, use fittings with a very pronounced, gradual sweep. Common fittings for this purpose include the combination Y-fitting with 45° elbow (often called a *combo*), and long sweep 90° elbow.

Horizontal offsets in drains: Y-fittings, 45° elbows, 22° elbows, and long sweep 90° elbows are used when changing directions in horizontal pipe runs. Whenever possible, horizontal drain pipes should use gradual, sweeping bends rather than sharp turns.

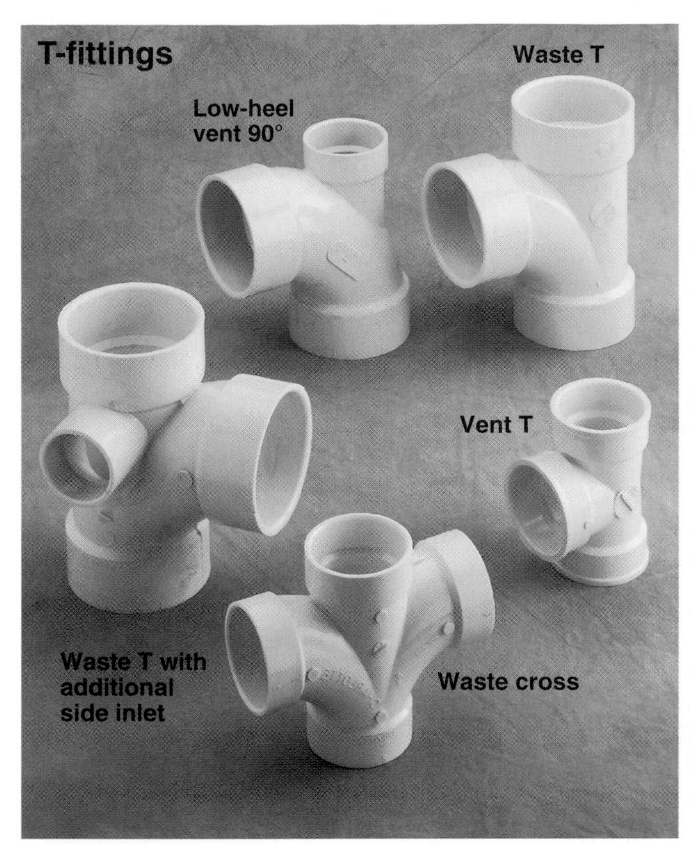

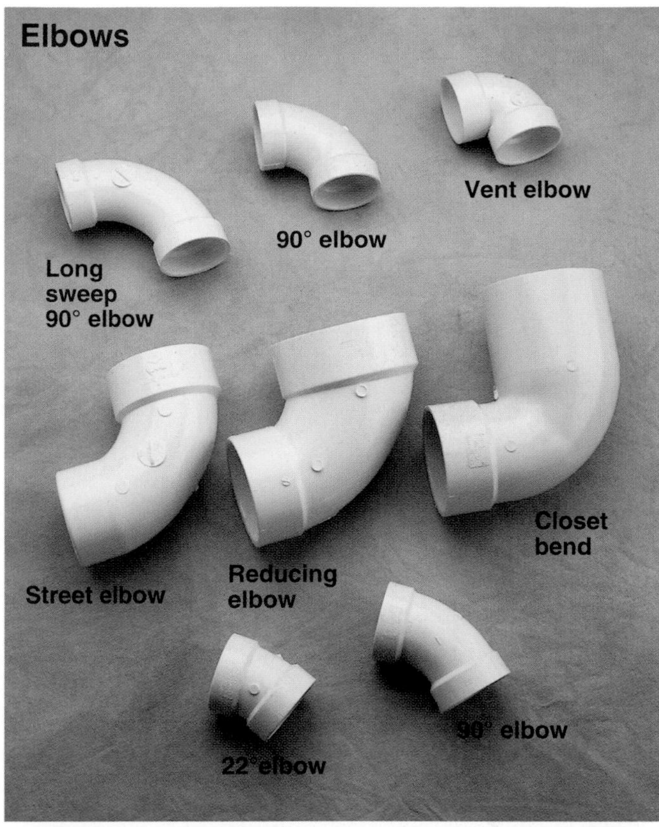

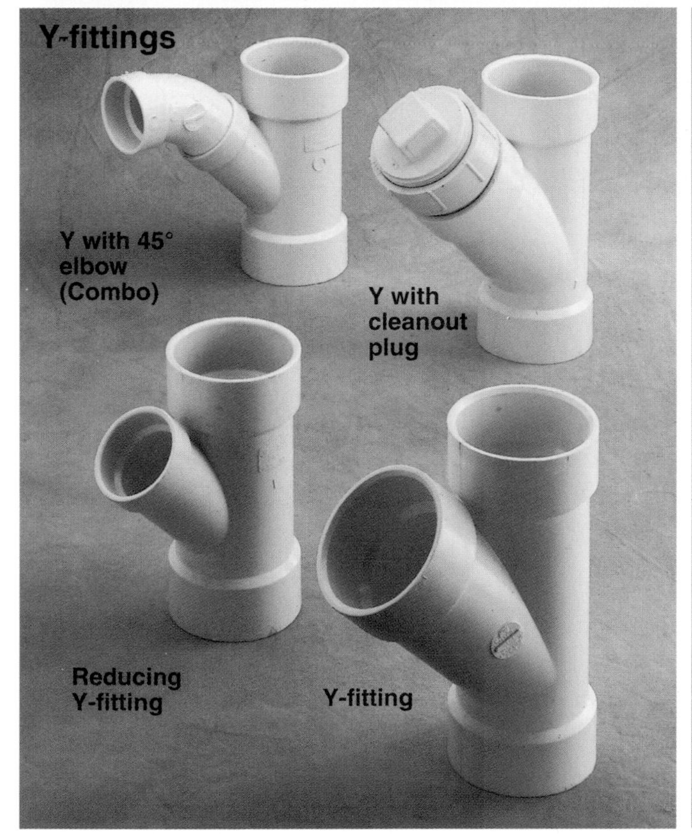

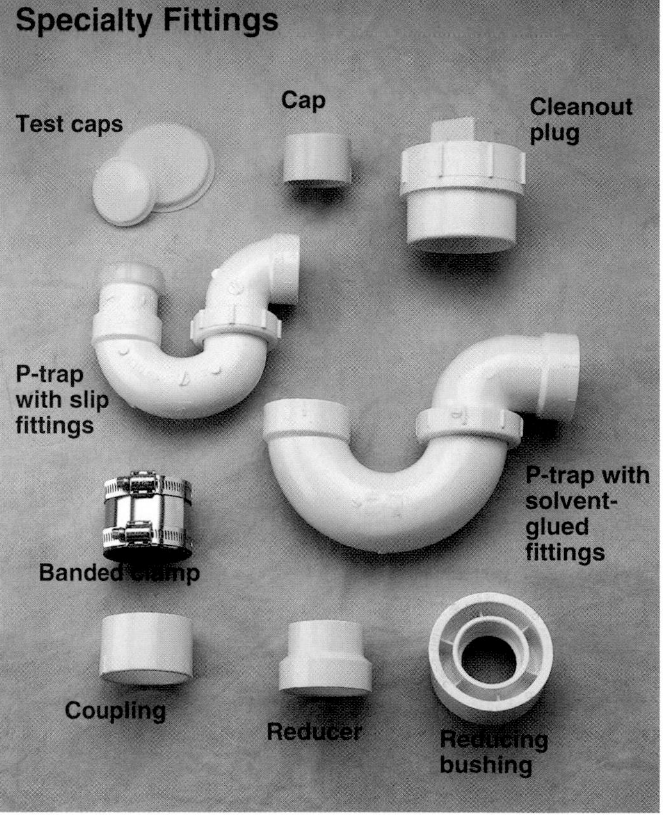

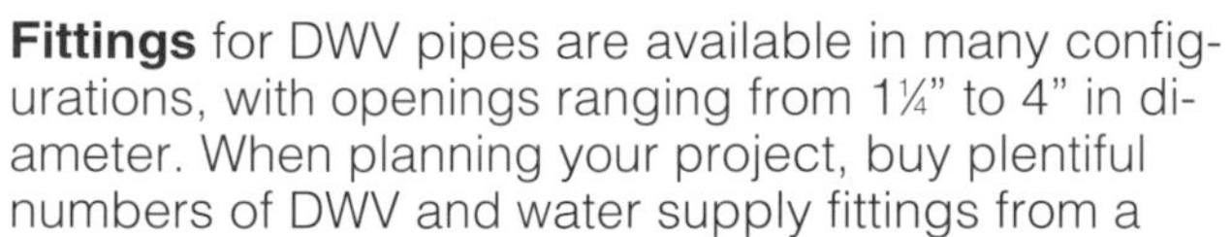

Fittings for DWV pipes are available in many configurations, with openings ranging from 1¼" to 4" in diameter. When planning your project, buy plentiful numbers of DWV and water supply fittings from a reputable retailer with a good return policy. It is much more efficient to return leftover materials after you complete your project than it is to interrupt your work each time you need to shop for a missing fitting.

The plumbing inspector is the final authority when it comes to evaluating your work. By visually examining and testing your new plumbing, the inspector ensures that your work is safe and functional.

Understanding Plumbing Codes

The Plumbing Code is the set of regulations that building officials and inspectors use to evaluate your project plans and the quality of your work. Codes vary from region to region, but most are based on the National Uniform Plumbing Code, the authority we used in the development of this book.

Code books are available for reference at bookstores and government offices. However, they are highly technical, difficult-to-read manuals. More user-friendly for do-it-yourselfers are the variety of Code Handbooks available at bookstores and libraries. These handbooks are based on the National Uniform Plumbing Code, but are easier to read and include many helpful diagrams and photos.

Plumbing Code Handbooks sometimes discuss three different plumbing "zones" in an effort to accommodate variations in regulations from state to state. The states included in each zone are listed below.

Zone 1: Washington, Oregon, California, Nevada, Idaho, Montana, Wyoming, North Dakota, South Dakota, Minnesota, Iowa, Nebraska, Kansas, Utah, Arizona, Colorado, New Mexico, Indiana, parts of Texas.

Zone 2: Alabama, Arkansas, Louisiana, Tennessee, North Carolina, Mississippi, Georgia, Florida, South Carolina, parts of Texas, parts of Maryland, parts of Delaware, parts of Oklahoma, parts of West Virginia.

Zone 3: Virginia, Kentucky, Missouri, Illinois, Michigan, Ohio, Pennsylvania, New York, Connecticut, Massachusetts, Vermont, New Hampshire, Rhode Island, New Jersey, parts of Delaware, parts of West Virginia, parts of Maine, parts of Maryland, parts of Oklahoma.

Remember that your local Plumbing Code always supercedes the National Code. On some issues, the local Code may be less demanding than the National Code, but on other issues it may be more restrictive. Your local building inspector is a valuable source of information and may provide you with a convenient summary sheet of the regulations that apply to your project.

Getting a Permit

To ensure public safety, your community requires that you obtain a permit for most plumbing projects, including all the projects demonstrated in this book.

When you visit your city Building Inspection office to apply for a permit, the building official will want to review three drawings of your plumbing project: a site plan, a water supply diagram, and a drain-waste-vent diagram. These drawings are described on this page. If the official is satisfied that your project meets Code requirements, he will issue you a plumbing permit, which is your legal permission to begin work. The building official will also specify an inspection schedule for your project. As your project nears completion, you will be asked to arrange for an inspector to visit your home while the pipes are exposed and review the installation to ensure its safety.

Although do-it-yourselfers often complete complex plumbing projects without obtaining a permit or having the work inspected, we strongly urge you to comply with the legal requirements in your area. A flawed plumbing system can be dangerous, and it can potentially threaten the value of your home.

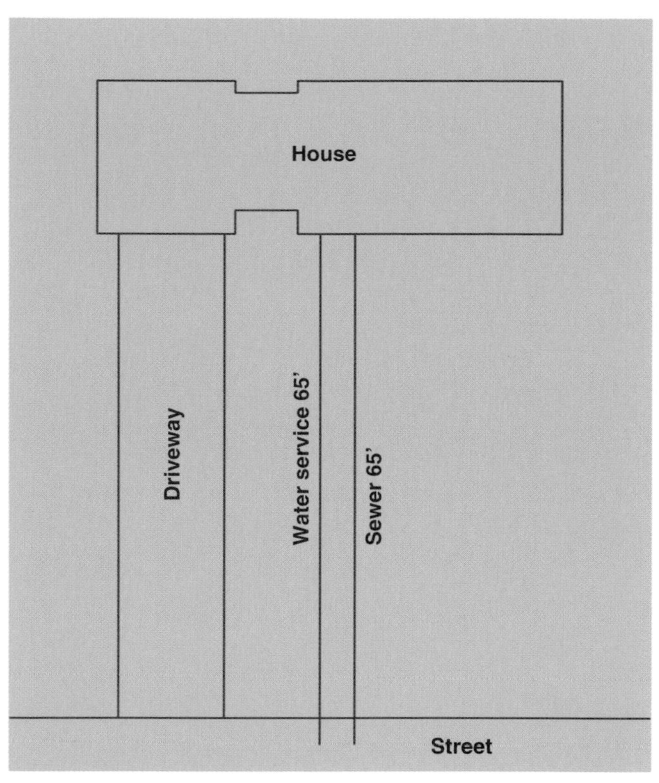

The site plan shows the location of the water main and sewer main with respect to your yard and home. The distances from your foundation to the water main and from the foundation to the main sewer should be indicated on the site plan.

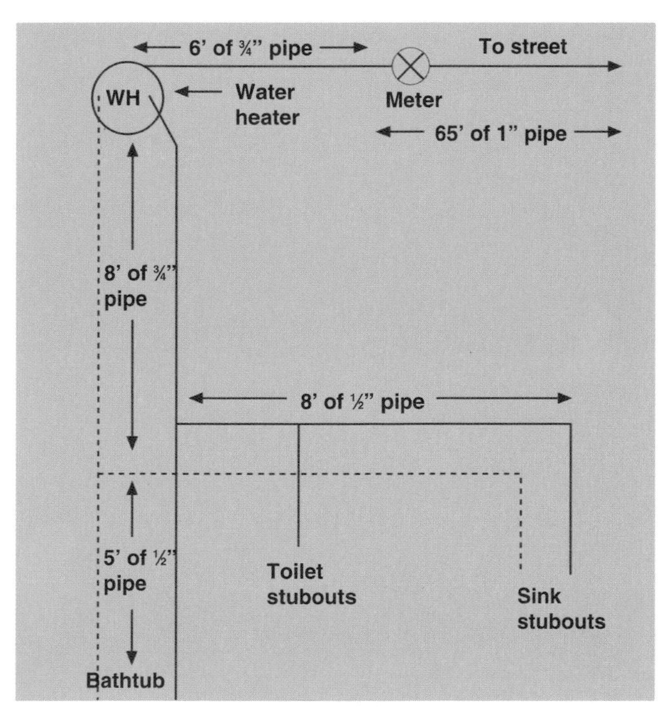

The supply riser diagram shows the length of the hot and cold water pipes and the relation of the fixtures to one another. The inspector will use this diagram to determine the proper size for the new water supply pipes in your system.

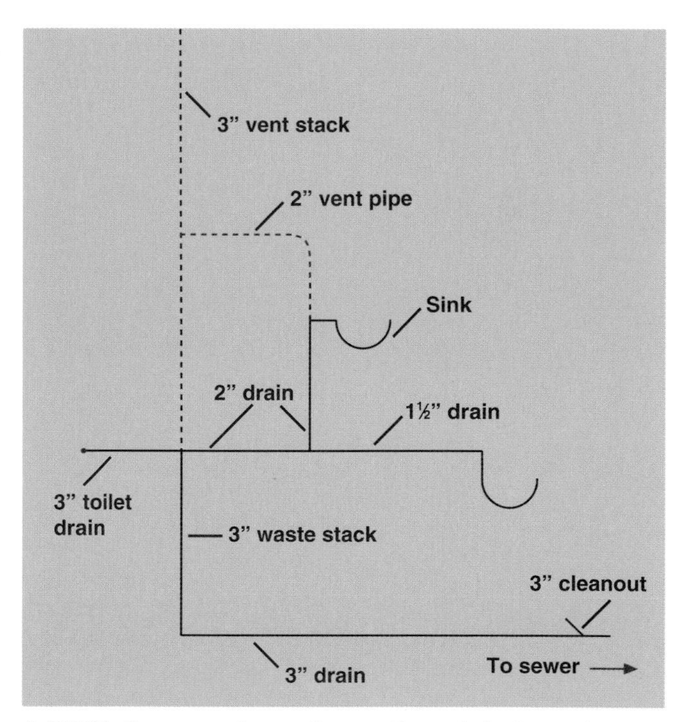

A DWV diagram shows the routing of drain and vent pipes in your system. Make sure to indicate the lengths of drain pipes and the distances between fixtures. The inspector will use this diagram to determine if you have properly sized the drain traps, drain pipes, and vent pipes in your project.

Sizing for Water Distribution Pipes

Fixture	Unit rating
Toilet	3
Vanity sink	1
Shower	2
Bathtub	2
Dishwasher	2
Kitchen sink	2
Clothes washer	2
Utility sink	2
Sillcock	3

Size of service pipe from street	Size of distribution pipe from water meter	Maximum length (ft.)—total fixture units					
		40	60	80	100	150	200
¾"	½"	9	8	7	6	5	4
¾"	¾"	27	23	19	17	14	11
¾"	1"	44	40	36	33	28	23
1"	1"	60	47	41	36	30	25
1"	1¼"	102	87	76	67	52	44

Water distribution pipes are the main pipes extending from the water meter throughout the house, supplying water to the branch pipes leading to individual fixtures. To determine the size of the distribution pipes, you must first calculate the total demand in "fixture units" (above, left) and the overall length of the water supply lines, from the street hookup through the water meter and to the most distant fixture in the house. Then, use the second table (above, right) to calculate the minimum size for the water distribution pipes. Note that the fixture unit capacity depends partly on the size of the street-side pipe that delivers water to your meter.

Sizes for Branch Pipes & Supply Tubes

Fixture	Min. branch pipe size	Min. supply tube size
Toilet	½"	⅜"
Vanity sink	½"	⅜"
Shower	½"	½"
Bathtub	½"	½"
Dishwasher	½"	½"
Kitchen sink	½"	½"
Clothes washer	½"	½"
Utility sink	½"	½"
Sillcock	¾"	N.A.
Water heater	¾"	N.A.

Branch pipes are the water supply lines that run from the distribution pipes toward the individual fixtures. **Supply tubes** are the vinyl, chromed copper, or mesh tubes that carry water from the branch pipes to the fixtures. Use the chart above as a guide when sizing branch pipes and supply tubes.

Valve Requirements

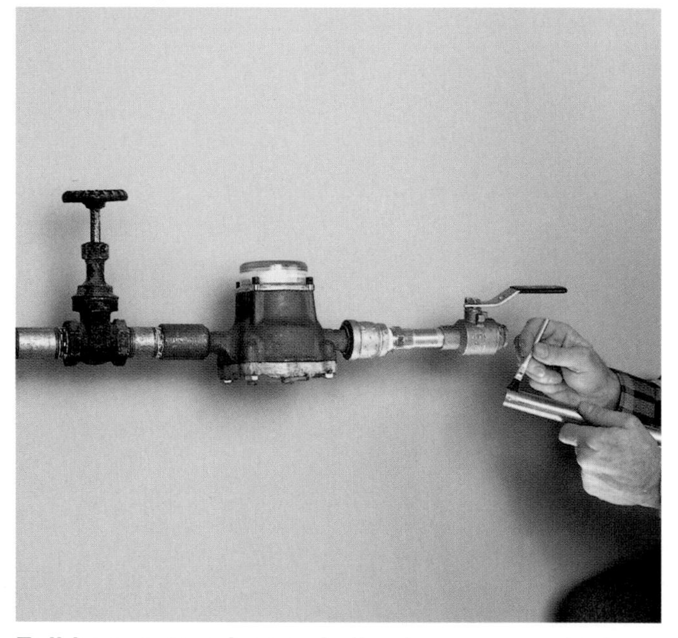

Full-bore gate valves or ball valves are required in the following locations: on both the street side and house side of the water meter; on the inlet pipes for water heaters and heating system boilers. Individual fixtures should have accessible shutoff valves, but these need not be full-bore valves. All sillcocks must have individual control valves located inside the house.

Modifying Water Pressure

Pressure-reducing valve (shown above) is required if the water pressure coming into your home is greater than 80 pounds per square inch (psi). The reducing valve should be installed near the point where the water service enters the building. A **booster pump** may be required if the water pressure in your home is below 40 psi.

Preventing Water Hammer

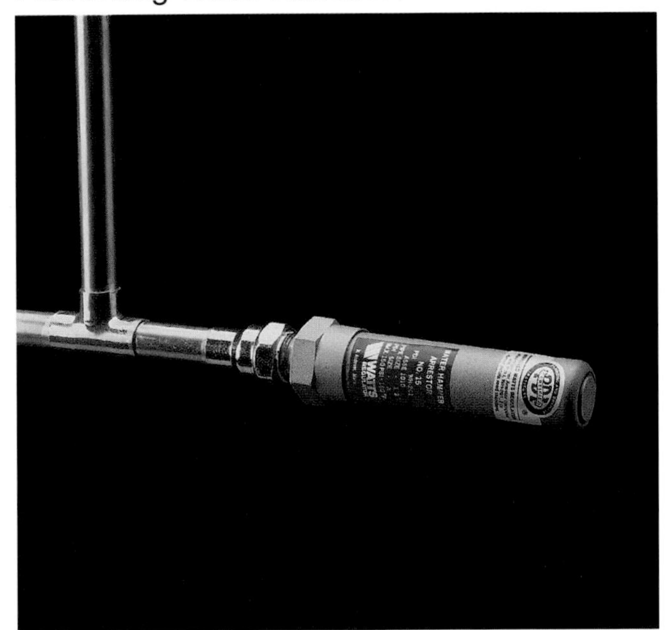

Water hammer arresters may be required by Code. Water hammer is a problem that may occur when the fast-acting valves on washing machines or other appliances cause pipes to vibrate against framing members. The arrester works as a shock absorber, with a watertight diaphragm inside. It is mounted to a T-fitting installed near the appliance.

Anti-syphon Devices

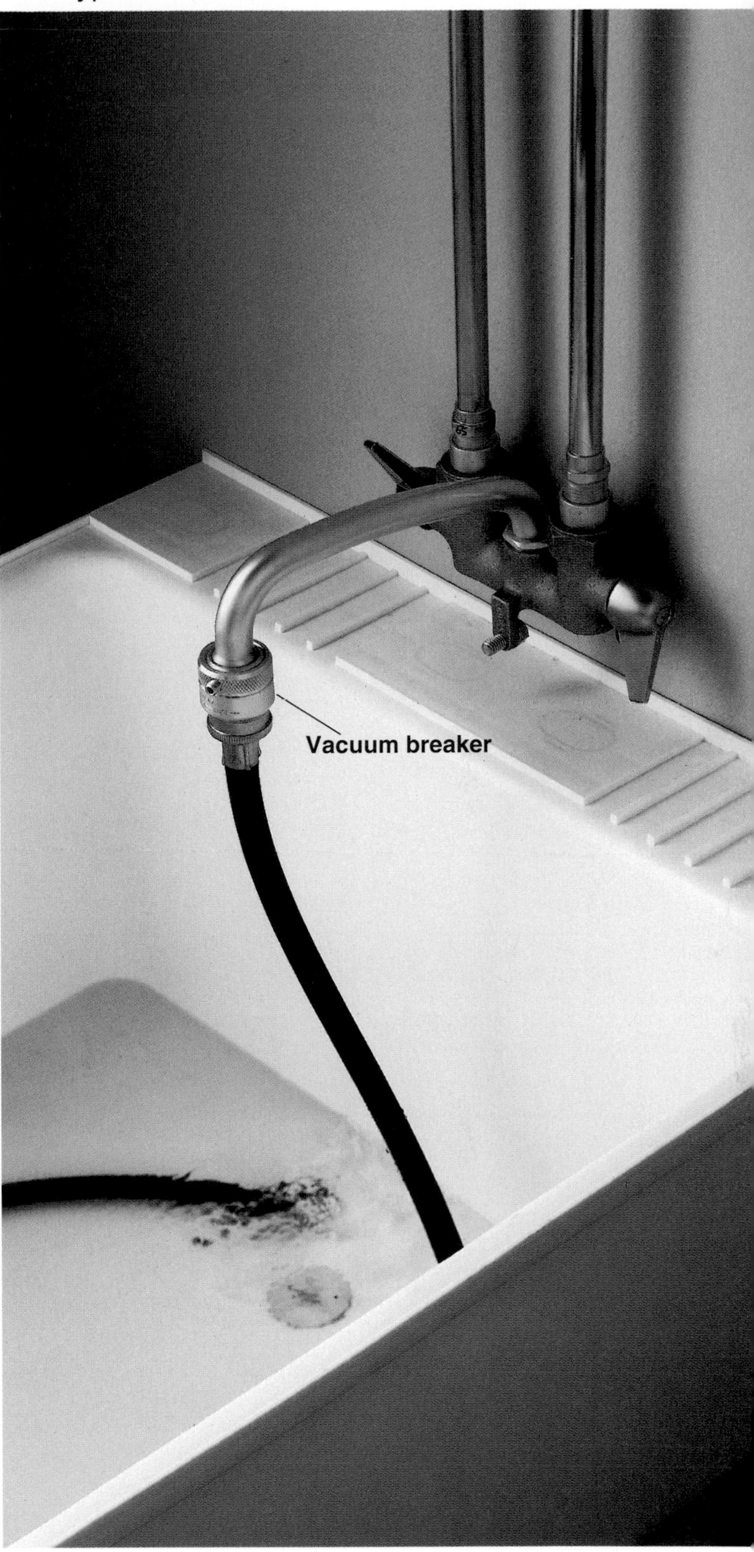

Vacuum breakers must be installed on all indoor and outdoor hose bibs and any outdoor branch pipes that run underground (page 131, step 7). Vacuum breakers prevent contaminated water from being drawn into the water supply pipes in the event of a sudden drop in water pressure in the water main. When a drop in pressure produces a partial vacuum, the breaker prevents siphoning by allowing air to enter the pipes.

Drain cleanouts make your DWV system easier to service. In most areas, the plumbing Code requires that you place cleanouts at the end of every horizontal drain run. Where horizontal runs are not accessible, removable drain traps will suffice as cleanouts.

Pipe Support Intervals

Type of pipe	Vertical support interval	Horizontal support interval
Copper	6 ft.	10 ft.
ABS	4 ft.	4 ft.
CPVC	3 ft.	3 ft.
PVC	4 ft.	4 ft.
Steel	12 ft.	15 ft.
Iron	5 ft.	15 ft.

Minimum intervals for supporting pipes are determined by the type of pipe and its orientation in the system. See page 83 for acceptable pipe support materials. Remember that the measurements shown above are minimum requirements; many plumbers install pipe supports at closer intervals.

Fixture Units & Minimum Trap Size

Fixture	Fixture units	Min. trap size
Shower	2	2"
Vanity sink	1	1¼"
Bathtub	2	1½"
Dishwasher	2	1½"
Kitchen sink	2	1½"
Kitchen sink*	3	1½"
Clothes washer	2	1½"
Utility sink	2	1½"
Floor drain	1	2"

*Kitchen sink with attached food disposer

Minimum trap size for fixtures is determined by the drain fixture unit rating, a unit of measure assigned by the Plumbing Code. NOTE: Kitchen sinks rate 3 units if they include an attached food disposer, 2 units otherwise.

Sizes for Horizontal & Vertical Drain Pipes

Pipe size	Maximum fixture units for horizontal branch drain	Maximum fixture units for vertical drain stacks
1¼"	1	2
1½"	3	4
2"	6	10
2½"	12	20
3"	20	30
4"	160	240

Drain pipe sizes are determined by the load on the pipes, as measured by the total fixture units. Horizontal drain pipes less than 3" in diameter should slope ¼" per foot toward the main drain. Pipes 3" or more in diameter should slope ⅛" per foot. NOTE: Horizontal or vertical drain pipes for a toilet must be 3" or larger.

Vent Pipe Sizes, Critical Distances

Size of fixture drain	Minimum vent pipe size	Maximum trap-to-vent distance
1¼"	1¼"	2½ ft.
1½"	1¼"	3½ ft.
2"	1½"	5 ft.
3"	2"	6 ft.
4"	3"	10 ft.

Vent pipes are usually one pipe size smaller than the drain pipes they serve. Code requires that the distance between the drain trap and the vent pipe fall within a maximum "critical distance," a measurement that is determined by the size of the fixture drain. Use this chart to determine both the minimum size for the vent pipe and the maximum critical distance.

Vent Pipe Orientation to Drain Pipe

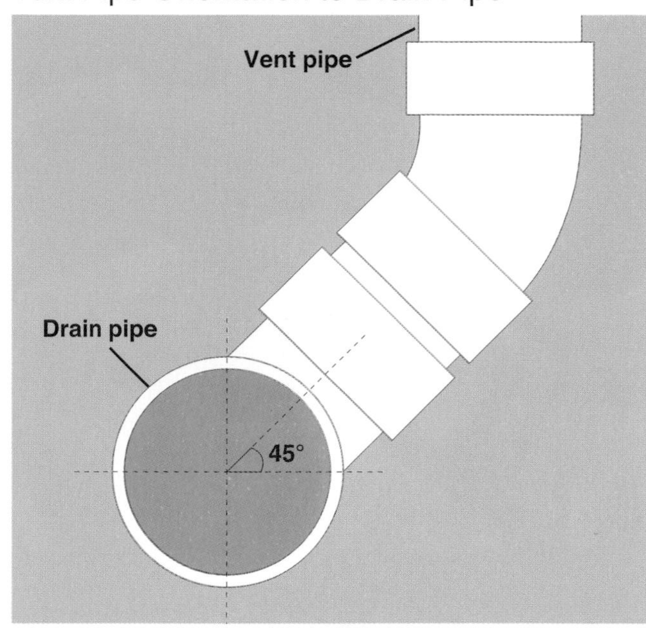

Vent pipes must extend in an upward direction from drains, no less than 45° from horizontal, This ensures that waste water cannot flow into the vent pipe and block it. At the opposite end, a new vent pipe should connect to an existing vent pipe or main waste-vent stack at a point at least 6" above the highest fixture draining into the system.

Wet Venting

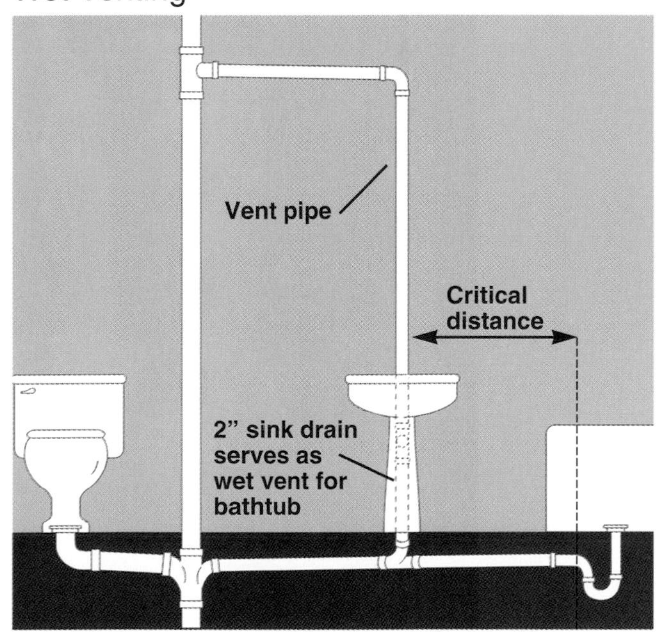

Wet vents are pipes that serve as a vent for one fixture and a drain for another. The sizing of a wet vent is based on the total fixture units it supports (opposite page): a 3" wet vent can serve up to 12 fixture units; a 2" wet vent is rated for 4 fixture units; a 1½" wet vent, for only 1 fixture unit. NOTE: The distance between the wet-vented fixture and the wet vent itself must be no more than the maximum critical distance (above, left).

Auxiliary Venting

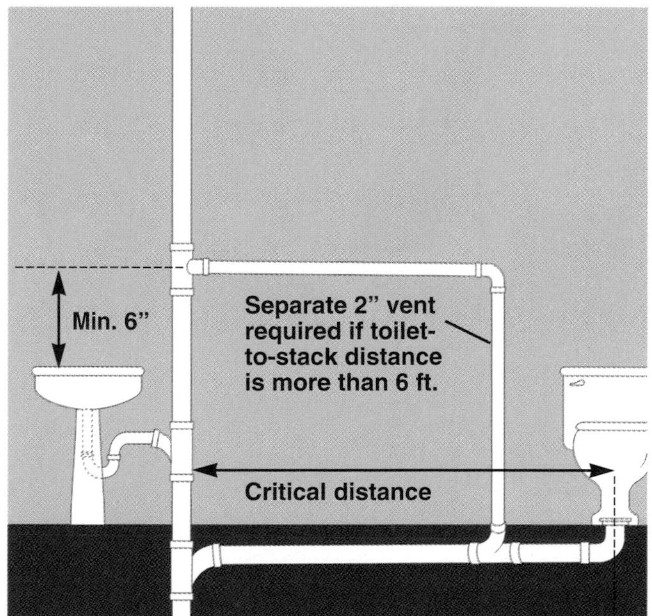

Fixtures must have auxiliary vents if the distance to the main waste-vent stack exceeds the critical distance (above, left). A toilet, for example, should have a separate vent pipe if it is located more than 6 ft. from the main waste-vent stack. This secondary vent pipe should connect to the stack or an existing vent pipe at a point at least 6" above the highest fixture on the system.

Testing New Plumbing Pipes

A **pressure gauge and air pump** are used to test DWV lines. The system is first blocked off at each fixture and at points near where the new drain and vent pipes connect to the main stack. Air is then pumped into the system to a pressure of 5 pounds per square inch (PSI). To pass inspection, the system must hold this pressure for 15 minutes.

When the building inspector comes to review your new plumbing, he may require that you perform a pressure test on the DWV and water supply lines as he watches. The inspection and test should be performed after the system is completed, but before the new pipes are covered with wallboard. To ensure that the inspection goes smoothly, it is a good idea to perform your own pretest, so you can locate and repair any problems before the inspector makes his visit.

The DWV system is tested by blocking off the new drain and vent pipes, then pressuring the system with air to see if it leaks. At the fixture stub-outs, the DWV pipes can be capped off or plugged with test balloons designed for this purpose. The air pump, pressure gauge, and test balloons required to test the DWV system can be obtained at tool rental centers.

Testing the water supply lines is a simple matter of turning on the water and examining the joints for leaks. If you find a leak, you will need to drain the pipes, then resolder the faulty joints.

How to Test New DWV Pipes

1 Insert a test balloon into the test T-fittings at the top and bottom of the new DWV line, blocking the pipes entirely. NOTE: Ordinary T-fittings installed near the bottom of the drain line and near the top of the vent line are generally used for test fittings.

2 Block toilet drains with a test balloon designed for a toilet bend. Large test balloons may need to be inflated with an air pump.

3 Cap off the remaining fixture drains by solvent-gluing test caps onto the stub-outs. After the system is tested, these caps are simply knocked loose with a hammer.

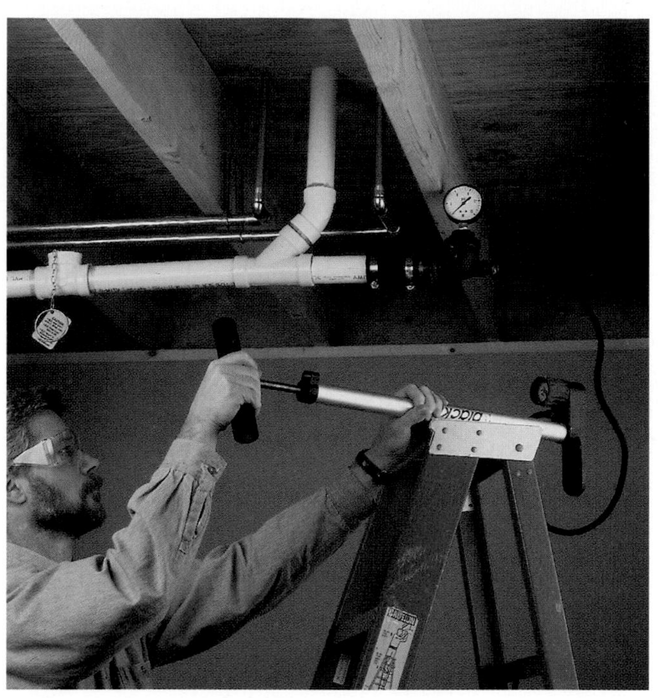

4 At a cleanout fitting, insert a *weenie*—a special test balloon with an air gauge and inflation valve. Attach an air pump to the valve on the weenie, and pressurize the pipes to 5 psi. Watch the pressure gauge for 15 minutes to ensure that the system does not lose pressure.

5 If the DWV system loses air when pressurized, check each joint for leaks by rubbing soapy water over the fittings and looking for active bubbles. When you identify a problem joint, cut away the existing fitting and solvent-glue a new fitting in place, using couplings and short lengths of pipe.

6 After the DWV system has been inspected and approved by a building official, remove the test balloons and close the test T-fittings by solvent-gluing caps onto the open inlets.

Use 2 x 6 studs to frame "wet walls" when constructing a new bathroom or kitchen. Thicker walls provide more room to run drain pipes and main waste-vent stacks, making installation much easier.

Installing New Plumbing

A major plumbing project is a complicated affair that often requires demolition and carpentry skills. Bathroom or kitchen plumbing may be unusable for several days while completing the work, so make sure you have a backup bathroom or kitchen space to use during this time.

To ensure that your project goes quickly, always buy plenty of pipe and fittings—at least 25% more than you think you need. Making several extra trips to the building center for last-minute fittings is a nuisance, and it can add many hours of time to your project. Always purchase from a reputable retailer that will allow you to return leftover fittings for credit.

The how-to projects on the following pages demonstrate standard plumbing techniques, but should not be used as a literal blueprint for your own work. Pipe and fitting sizes, fixture layout, and pipe routing will always vary according to individual circumstances. When planning your project, carefully read all the information in the Planning section, especially the material on Understanding Plumbing Codes (pages 86 to 91). Before you begin work, create a detailed plumbing plan to guide your work and help you obtain the required permits. This section includes information on:

- Plumbing Bathrooms (pages 98 to 115)
- Plumbing a Kitchen (pages 116 to 127)
- Installing Outdoor Plumbing (pages 128 to 133)

Tips for Installing New Plumbing

Use masking tape to mark the locations of fixtures and pipes on the walls and floors. Read the layout specifications that come with each sink, tub, or toilet, then mark the drain and supply lines accordingly. Position the fixtures on the floor, and outline them with tape. Measure and adjust until the arrangement is comfortable to you and meets minimum clearance specifications. If you are working in a finished room, prevent damage to wallpaper or paint by using self-adhesive notes to mark the walls.

Consider the location of cabinets when roughing in the water supply and drain stub-outs. You may want to temporarily position the cabinets in their final locations before completing the drain and water supply runs.

Install control valves at the points where the new branch supply lines meet the main distribution pipes. By installing valves, you can continue to supply the rest of the house with water while you are working on the new branches.

(continued next page)

Tips for Installing New Plumbing (continued)

Framing Member	Maximum Hole Size	Maximum Notch Size
2 × 4 loadbearing stud	1 7/16" diameter	7/8" deep
2 × 4 non-loadbearing stud	2 1/2" diameter	1 7/16" deep
2 × 6 loadbearing stud	2 1/4" diameter	1 3/8" deep
2 × 6 non-loadbearing stud	3 5/16" diameter	2 3/16" deep
2 × 6 joists	1 1/2" diameter	7/8" deep
2 × 8 joists	2 3/8" diameter	1 1/4" deep
2 × 10 joists	3 1/16" diameter	1 1/2" deep
2 × 12 joists	3 3/4" diameter	1 7/8" deep

Framing member chart shows the maximum sizes for holes and notches that can be cut into studs and joists when running pipes. Where possible, use notches rather than bored holes, because pipe installation is usually easer. When boring holes, there must be at least 5/8" of wood between the edge of a stud and the hole, and at least 2" between the edge of a joist and the hole. Joists can be notched only in the end one-third of the overall span; never in the middle one-third of the joist. When two pipes are run through a stud, the pipes should be stacked one over the other, never side by side.

Create access panels so that in the future you will be able to service fixture fittings and shutoff valves located inside the walls. Frame an opening between studs, then trim the opening with wood moldings. Cover the opening with a removable plywood panel the same thickness as the wall surface, then finish it to match the surrounding walls.

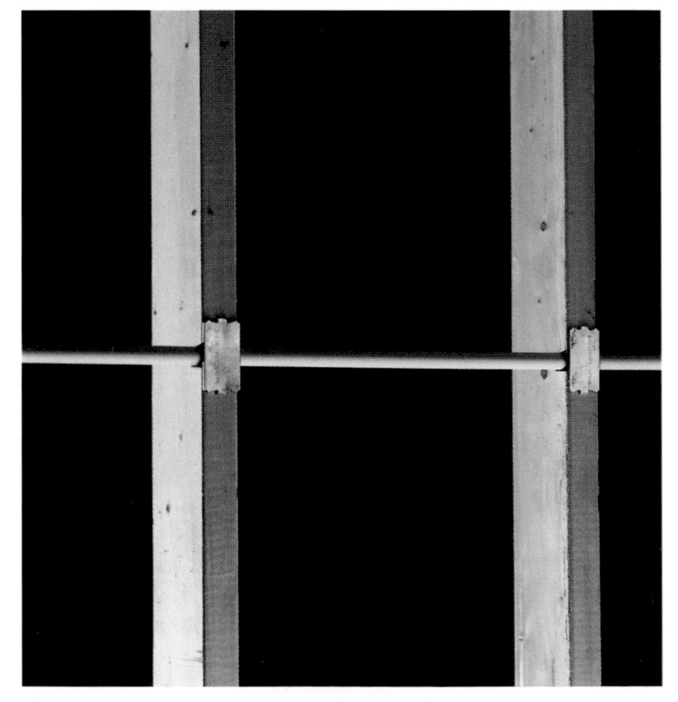

Protect pipes from punctures, if they are less than 1 1/4" from the front face of wall studs or joists, by attaching metal protector plates to the framing members.

Test-fit materials before solvent-gluing or soldering joints. Test-fitting ensures that you have the correct fittings and enough pipe to do the job, and can help you avoid lengthy delays during installation.

Support pipes adequately. Horizontal and vertical runs of DWV and water supply pipe must be supported at minimum intervals, which are specified by your local Plumbing Code (page 90). A variety of metal and plastic materials are available for supporting plumbing pipes (page 83).

Use plastic bushings to help hold plumbing pipes securely in holes bored through wall plates, studs, and joists. Bushings can help to cushion the pipes, preventing wear and reducing rattling.

Install extra T-fittings on new drain and vent lines so that you can pressure-test the system when the building inspector reviews your installation (pages 92 to 93). A new DWV line should have these extra T-fittings near the points where the new branch drains and vent pipes reach the main waste-vent stack.

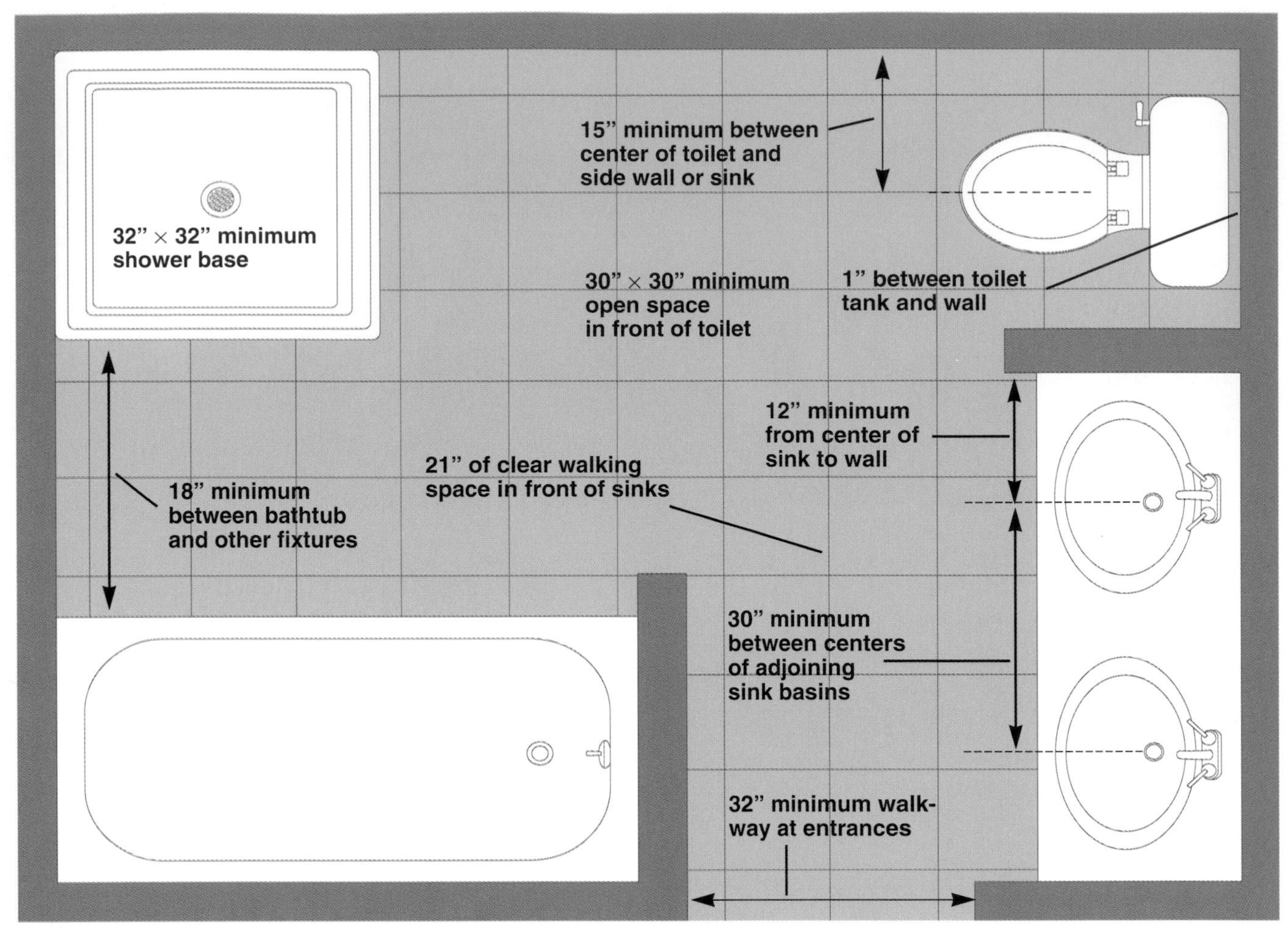

Follow minimum clearance guidelines when planning the locations of bathroom fixtures. Unobstructed access to the fixtures is fundamental to creating a comfortable, safe, and efficient bathroom.

Plumbing Bathrooms

Adding a new bathroom or updating an old one is a sure way to add value to your home. The personal comfort gained from a custom master bathroom can add a new dimension to your relaxation time. And adding a full bath in the basement or a half bath next to the kitchen offers convenience for both family members and guests. When planned and built correctly, a new or remodeled bathroom also improves the resale value of your home.

The first step when planning a bathroom is determining the type of bathroom you want. Do you have your heart set on expanding into a spare bedroom to create the ultimate master bathroom, or do you simply need a functional half bath for convenience? In this section, you'll see three demonstration projects that represent the full range of bathrooms, from a spacious master bathroom to a simple half bath (opposite page).

Next, you'll need to decide on the type of fixtures you need. Visit your local building centers to determine the range of fixtures available and their prices, and visit model homes and remodeling exhibitions to see how professionals arrange and install the fixtures you plan to use.

Once you have determined the scope of your project and settled on a budget, you can develop working plans for your bathroom. When creating a floor plan, always follow minimum clearance guidelines (above), and think about where the drain and water supply pipes will run. You can save yourself many hours of work by positioning fixtures so the pipes can be routed with simple, straight runs rather than with many complicated bends. Make sure that your project complies with local Plumbing Code regulations for bathroom plumbing.

Master bathroom can include luxury features, such as a large whirlpool tub or multi-jet shower. Our project contains both these features, as well as a pedestal vanity sink and toilet. A spacious bathroom may require substantial construction work if you intend to expand into an adjoining room. See pages 100 to 107.

Basement bathroom is ideal if you have bedrooms or finished recreation areas in your basement. Our project includes a shower, toilet, and vanity sink. Plumbing a basement bathroom may require that you break into the concrete floor to connect drain pipes. See pages 108 to 113.

Half bath can be easily added to a room that shares a "wet wall" with a kitchen or other bathroom. Our project includes a toilet and vanity sink. See pages 114 to 115.

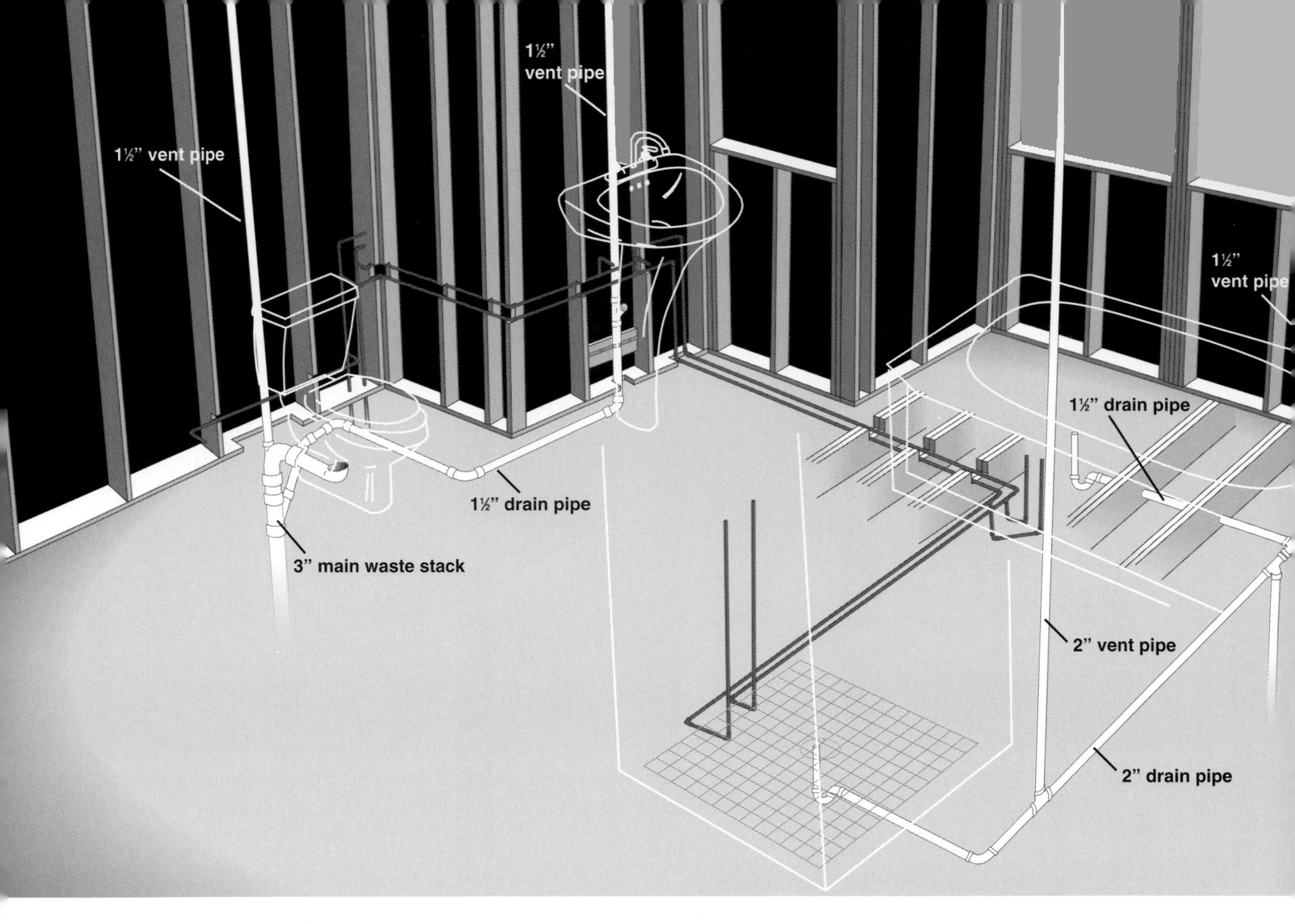

Plumbing a Master Bathroom

A large bathroom has more plumbing fixtures and consumes more water than any other room in your house. For this reason, a master bath has special plumbing needs.

Frame bathroom "wet walls" with 2 × 6 studs, to provide plenty of room for running 3" pipes and fittings. If your bathroom includes a heavy whirlpool tub, you will need to strengthen the floor by installing "sister" joists alongside the existing floor joists underneath the tub.

For convenience, our project is divided into the following sequences:

- How to Install DWV Pipes for the Toilet & Sink (pages 101 to 103)
- How to Install DWV Pipes for the Tub & Shower (pages 104 to 105)
- How to Connect the Drain Pipes & Vent Pipes to the Main Waste-Vent Stack (page 106)
- How to Install the Water Supply Pipes (page 107)

Our demonstration bathroom is a second-story master bath. We are installing a 3" vertical drain pipe to service the toilet and the vanity sink, and a 2" vertical pipe to handle the tub and shower drains. The branch drains for the sink and bathtub are 1½" pipes; for the shower, 2" pipe. Each fixture has its own vent pipe extending up into the attic, where they are joined together and connected to the main stack.

How to Install DWV Pipes for the Toilet & Sink

1 Use masking tape to outline the locations of the fixtures and pipe runs on the subfloor and walls. Mark the location for a 3" vertical drain pipe on the sole plate in the wall behind the toilet. Mark a 4½"-diameter circle for the toilet drain on the subfloor.

2 Cut out the drain opening for the toilet, using a jig saw. Mark and remove a section of flooring around the toilet area, large enough to provide access for installing the toilet drain and for running drain pipe from the sink. Use a circular saw with blade set to the thickness of the flooring to cut through the subfloor.

3 If a floor joist interferes with the toilet drain, cut away a short section of the joist and box-frame the area with double headers.The framed opening should be just large enough to install the toilet and sink drains.

4 To create a path for the vertical 3" drain pipe, cut a 4½" × 12" notch in the sole plate of the wall behind the toilet. Make a similar cutout in the double wall plate at the bottom of the joist cavity. From the basement, locate the point directly below the cutout by measuring from a reference point, such as the main waste-vent stack.

5 Mark the location for the 3" drain pipe on the basement ceiling, then drill a 1"-diameter hole up through the center of the marked area. Direct the beam of a bright flashlight up into the hole, then return to the bathroom and look down into the wall cavity. If you can see light, return to the basement and cut a 4½"-diameter hole centered over the test hole.

(continued next page)

6 Measure and cut a length of 3" drain pipe to reach from the bathroom floor cavity to a point flush with the bottom of the ceiling joists in the basement. Solvent-glue a 3" × 3" × 1½" Y-fitting to the top of the pipe, and a low-heel vent 90° fitting above the Y. The branch inlet on the Y should face toward the sink location; the front inlet on the low-heel should face forward. Carefully lower the pipe into the wall cavity.

7 Lower the pipe so the bottom end slides through the opening in the basement ceiling. Support the pipe with vinyl pipe strap wrapped around the low-heel vent 90° fitting and screwed to framing members.

8 Use a length of 3" pipe and a 4" × 3" reducing elbow to extend the drain out to the toilet location. Make sure the drain slopes at least ⅛" per foot toward the wall, then support it with pipe strap attached to the joists. Insert a short length of pipe into the elbow, so it extends at least 2" above the subfloor. After the new drains are pressure tested, this stub-out will be cut flush with the subfloor and fitted with a toilet flange.

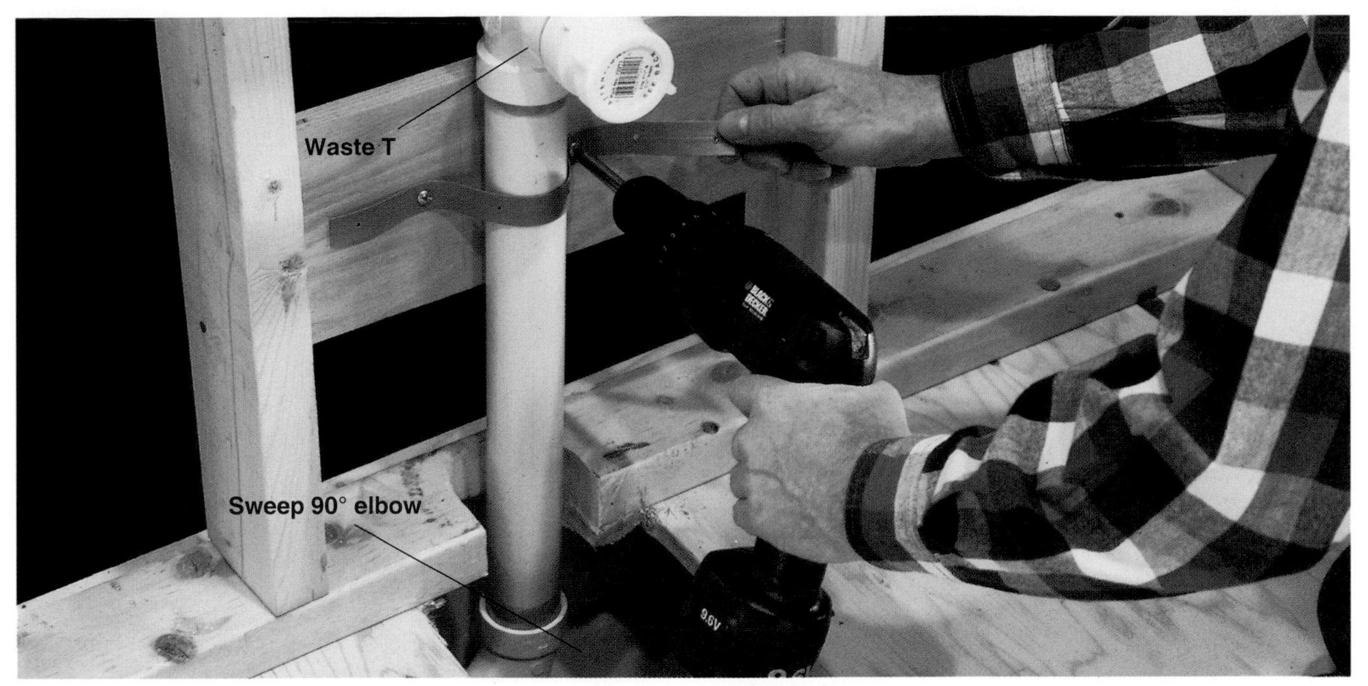

9 Notch out the sole plate and subfloor below the sink location. Cut a length of 1½" plastic drain pipe, then solvent-glue a waste T to the top of the pipe and a sweep 90° elbow to the bottom. The branch of the T should face out, and the discharge on the elbow should face toward the toilet location. Adjust the pipe so the top edge of the elbow nearly touches the bottom of the sole plate. Anchor it with a ¾"-thick backing board nailed between the studs.

10 Dry-fit lengths of 1½" drain pipe and elbows to extend the sink drain to the 3" drain pipe behind the toilet. Use a right-angle drill to bore holes in joists, if needed. Make sure the horizontal drain pipe slopes at least ¼" per foot toward the vertical drain. When satisfied with the layout, solvent-glue the pieces together and support the drain pipe with vinyl pipe straps attached to the joists.

11 In the top plates of the walls behind the sink and toilet, bore ½"-diameter holes up into the attic. Insert pencils or dowels into the holes, and tape them in place. Enter the attic and locate the pencils, then clear away insulation and cut 2"-diameter holes for the vertical vent pipes. Cut and install 1½" vent pipes running from the toilet and sink drain at least 1 ft. up into the attic.

How to Install DWV Pipes for the Tub & Shower

1 On the subfloor, use masking tape to mark the locations of the tub and shower, the water supply pipes, and the tub and shower drains, according to your plumbing plan. Use a jig saw to cut out a 12"-square opening for each drain, and drill 1"-diameter holes in the subfloor for each water supply riser.

2 When installing a large whirlpool tub, cut away the subfloor to expose the full length of the joists under the tub, then screw or bolt a second joist, called a *sister,* against each existing joist. Make sure both ends of each joist are supported by loadbearing walls.

3 In a wall adjacent to the tub, establish a route for a 2" vertical waste-vent pipe running from basement to attic.This pipe should be no more than 3½ ft. from the bathtub trap. Then, mark a route for the horizontal drain pipe running from the bathtub drain to the waste-vent pipe location. Cut 3"-diameter holes through the centers of the joists for the bathtub drain.

4 Cut and install a vertical 2" drain pipe running from basement to the joist cavity adjoining the tub location, using the same technique as for the toilet drain (steps 4 to 6, pages 101 to 102). At the top of the drain pipe, use assorted fittings to create three inlets: branch inlets for the bathtub and shower drains, and a 1½" top inlet for a vent pipe running to the attic.

5 Dry-fit a 1½" drain pipe running from the bathtub drain location to the vertical waste-vent pipe in the wall. Make sure the pipe slopes ¼" per foot toward the wall. When satisfied with the layout, solvent-glue the pieces together and support the pipe with vinyl pipe straps attached to the joists.

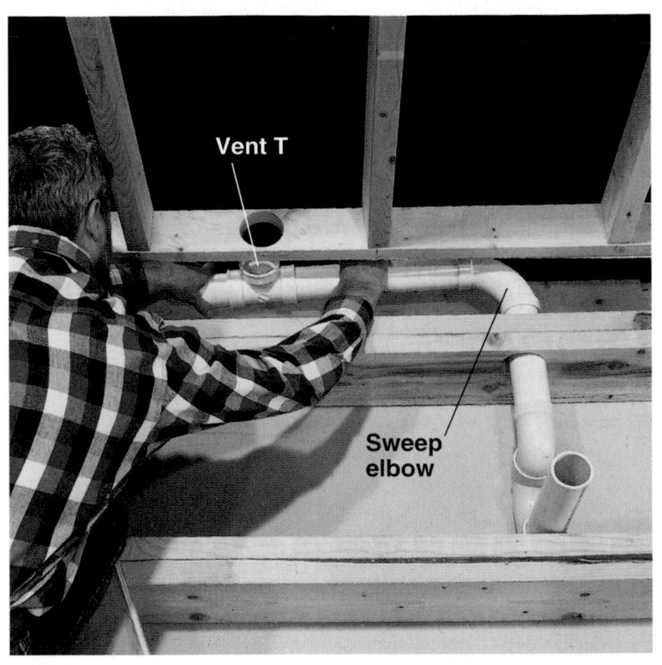

6 Dry-fit a 2" drain pipe from the shower drain to the vertical waste-vent pipe near the tub. Install a solvent-glued trap at the drain location, and cut a hole in the sole plate and insert a 2" × 2" × 1½" vent T within 5 ft. of the trap. Make sure the drain is sloped ¼" per foot downward away from the shower drain. When satisfied with the layout, solvent-glue the pipes together.

7 Cut and install vertical vent pipes for the bathtub and shower, extending up through the wall plates and at least 1 ft. into the attic. These vent pipes will be connected in the attic to the main waste-vent stack. In our project, the shower vent is a 2" pipe, while the bathtub vent is a 1½" pipe.

How to Connect the Drain Pipes to the Main Waste-Vent Stack

1 In the basement, cut into the main waste-vent stack and install the fittings necessary to connect the 3" toilet-sink drain and the 2" bathtub-shower drain. In our project, we created an assembly made of a waste T-fitting with an extra side inlet and two short lengths of pipe, then inserted it into the existing waste-vent stack using banded couplings. Make sure the T-fittings are positioned so the drain pipes will have the proper downward slope toward the stack.

2 Dry-fit Y-fittings with 45° elbows onto the vertical 3" and 2" drain pipes. Position the horizontal drain pipes against the fittings, and mark them for cutting. When satisfied with the layout, solvent-glue the pipes together, then support the pipes every 4 ft. with vinyl pipe straps. Solvent-glue cleanout plugs on the open inlets on the Y-fittings.

How to Connect the Vent Pipes to the Main Waste-Vent Stack

1 In the attic, cut into the main waste-vent stack and install a vent T-fitting, using banded couplings. The side outlet on the vent T should face the new 2" vent pipe running down to the bathroom. Attach a test T-fitting to the vent T. NOTE: If your stack is cast iron, make sure to adequately support it before cutting into it.

2 Use elbows, vent T-fittings, reducers, and lengths of pipe as needed to link the new vent pipes to the test T-fitting on the main waste-vent stack. Vent pipes can be routed in many ways, but you should make sure the pipes have a slight downward angle to prevent moisture from collecting in the pipes. Support the pipes every 4 ft.

How to Install the Water Supply Pipes

1 After shutting off the water, cut into existing supply pipes and install T-fittings for new branch lines. Notch out studs and run copper pipes to the toilet and sink locations. Use an elbow and threaded female fitting to form the toilet stub-out. Once satisfied with the layout, solder the pipes in place.

2 Cut 1" × 4"-high notches around the wall, and extend the supply pipes to the sink location. Install reducing T-fittings and female threaded fittings for the sink faucet stub-outs. The stub-outs should be positioned about 18" above the floor, spaced 8" apart. Once satisfied with the layout, solder the joints, then insert ¾" blocking behind the stub-outs and strap them in place.

3 Extend the water supply pipes to the bathtub and shower. In our project, we removed the subfloor and notched the joists to run ¾" supply pipes from the sink to a whirlpool bathtub, then to the shower. At the bathtub, we used reducing T-fittings and elbows to create ½" risers for the tub faucet. Solder caps onto the risers; after the subfloor is replaced, the caps will be removed and replaced with shutoff valves.

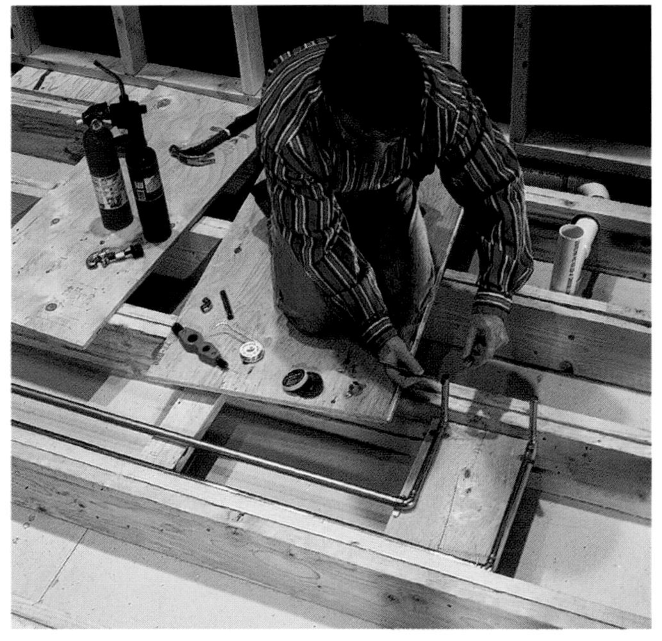

4 At the shower location, use elbows to create vertical risers where the shower wet wall will be constructed. The risers should extend at least 6" above floor level. Support the risers with a ¾" backer board attached between joists. Solder caps onto the risers. After the shower stall is constructed, the caps will be removed and replaced with shutoff valves.

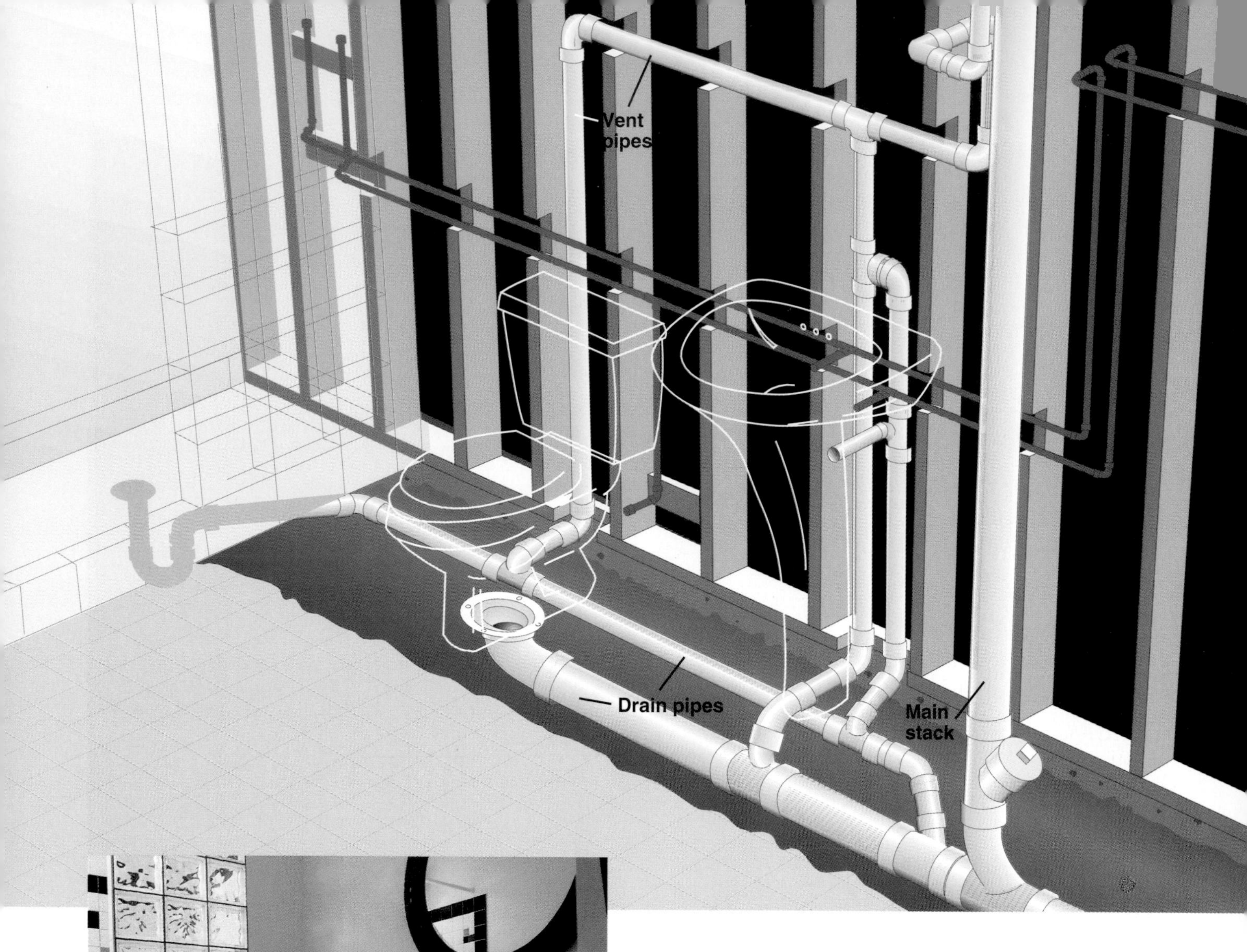

Our demonstration bathroom includes a shower, toilet, and vanity sink arranged in a line to simplify trenching. A 2" drain pipe services the new shower and sink; a 3" pipe services the new toilet. The drain pipes converge at a Y-fitting joined to the existing main drain. The shower, toilet, and sink have individual vent pipes that meet inside the wet wall before extending up into the attic, where they join the main waste-vent stack.

Plumbing a Basement Bath

When installing a basement bath, make sure you allow extra time for tearing out the concrete floor to accommodate the drains, and for construction of a wet wall to enclose supply and vent pipes. Constructing your wet wall with 2 × 6 studs and plates will provide ample room for running pipes. Be sure to schedule an inspection by a building official before you replace the concrete and cover the walls with wallboard.

Whenever possible, try to hold down costs by locating your basement bath close to existing drains and supply pipes.

How to Plumb a Basement Bath

1 Outline a 24"-wide trench on the concrete where new branch drains will run to the main drain. In our project, we ran the trench parallel to an outside wall, leaving a 6" ledge for framing a wet wall. Use a masonry chisel and hand maul to break up concrete near the stack.

2 Use a circular saw and masonry blade to cut along the outline, then break the rest of the trench into convenient chunks with a jackhammer. Remove any remaining concrete with a chisel. Excavate the trench to a depth about 2" deeper than the main drain. At vent locations for the shower and toilet, cut 3" notches in the concrete all the way to the wall.

3 Cut the 2 × 6 framing for the wet wall that will hold the pipes. Cut 3" notches in the bottom plate for the pipes, then secure the plate to the floor with construction adhesive and masonry nails. Install the top plate, then attach studs.

4 Assemble a 2" horizontal drain pipe for the sink and shower, and a 3" drain pipe for the toilet. The 2" drain pipe includes a solvent-glued trap for the shower, a vent T, and a waste T for the sink drain. The toilet drain includes a toilet bend and a vent T. Use elbows and straight lengths of pipe to extend the vent and drain pipes to the wet wall. Make sure the vent fittings angle upward from the drain pipe at least 45°.

(continued next page)

How to Plumb a Basement Bath (continued)

5 Use pairs of stakes with vinyl support straps slung between them to cradle drain pipes in the proper position (inset). The drain pipes should be positioned so they slope ¼" per foot down toward the main drain.

6 Assemble the fittings required to tie the new branch drains into the main drain. In our project, we will be cutting out the cleanout and sweep on the main waste-vent stack in order to install a new assembly that includes a Y-fitting to accept the two new drain pipes.

7 Support the main waste-vent stack before cutting. Use a 2 × 4 for a plastic stack, or riser clamps (page 145) for a cast-iron stack. Using a reciprocating saw (or cast iron cutter), cut into the main drain as close as possible to the stack.

8 Cut into the stack above the cleanout and remove the pipe and fittings. Wear rubber gloves, and have a bucket and plastic bags ready, as old pipes and fittings may be coated with messy sludge.

9 Test-fit, then solvent-glue the new cleanout and reducing Y assembly into the main drain. Support the weight of the stack by adding sand underneath the Y, but leave plenty of space around the end for connecting the new branch pipes.

10 Working from the reducing Y, solvent-glue the new drain pipes together. Be careful to maintain proper slope of the drain pipes when gluing. Be sure the toilet and shower drains extend at least 2" above the floor level.

11 Check for leaks by pouring fresh water into each new drain pipe. If no leaks appear, cap or plug the drains with rags to prevent sewer gas from leaking into the work area as you complete the installation.

(continued next page)

12 Run 2" vent pipes from the drains up the inside of the wet wall. Notch the studs and insert a horizontal vent pipe, then attach the vertical vent pipes with an elbow and vent T-fitting. Test-fit all pipes, then solvent-glue them in place.

13 Route the vent pipe from the wet wall to a point below a wall cavity running from the basement to the attic. NOTE: If there is an existing vent pipe in the basement, you can tie into this pipe rather than run the vent to the attic.

14 If you are running vent pipes in a two-story home, remove sections of wall surface as needed to bore holes for running the vent pipe through wall plates. Feed the vent pipe up into the wall cavity from the basement.

15 Wedge the vent pipe in place while you solvent-glue the fittings. Support the vent pipe at each floor with vinyl pipe straps. Do not patch the walls until your work has been inspected by a building official.

16 Cut into the main stack in the attic, and install a vent T-fitting, using banded couplings. (If the stack is cast iron, make sure to support it adequately above and below the cuts.) Attach a test T-fitting to the vent T, then join the new vent pipe to the stack, using elbows and lengths of straight pipe as needed.

17 Shut off the main water supply, cut into the water supply pipes as near as possible to the new bathroom, and install T-fittings. Install full-bore control valves on each line, then run ¾" branch supply pipes down into the wet wall by notching the top wall plate. Extend the pipes across the wall by notching the studs.

18 Use reducing T-fittings to run ½" supplies to each fixture, ending with female threaded adapters. Install backing boards, and strap the pipes in place. Attach metal protector plates over notched studs to protect pipes. After having your work approved by a building official, fill in around the pipes with dirt or sand, then mix and pour new concrete to cover the trench. Trowel the surface smooth, and let the cement cure for 3 days before installing fixtures.

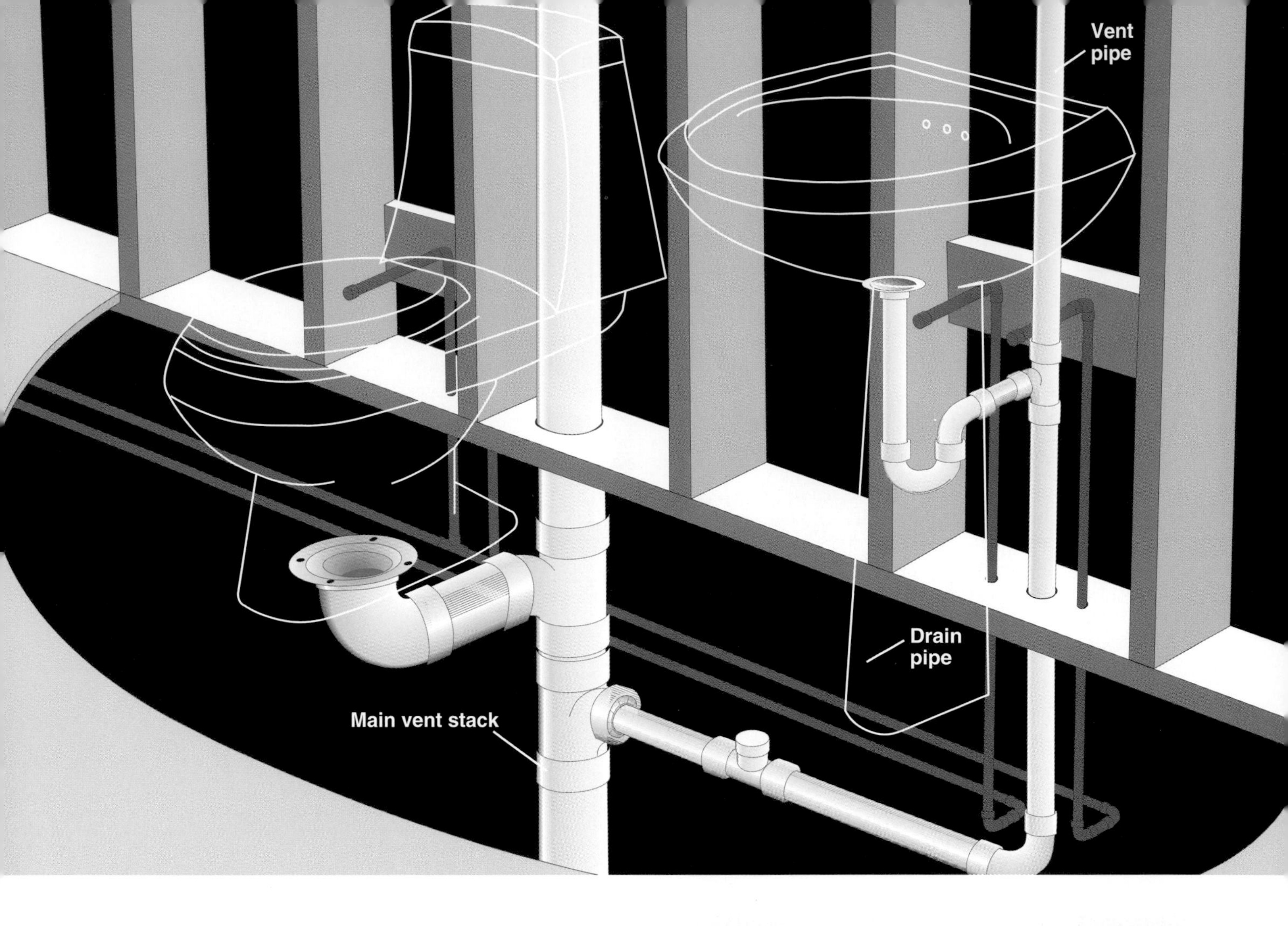

Plumbing a Half Bath

A first-story half bath is easy to install when located behind a kitchen or existing bathroom, because you can take advantage of accessible supply and DWV lines. It is possible to add a half bath on an upper story or in a location distant from existing plumbing, but the complexity and cost of the project may be increased considerably.

Be sure that the new fixtures are adequately vented. We vented the pedestal sink with a pipe that runs up the wall a few feet before turning to join the main stack. However, if there are higher fixtures draining into the main stack, you would be required to run the vent up to a point at least 6" above the highest fixture before splicing it into the main stack or an existing vent pipe. When the toilet is located within 6 ft. of the stack, as in our design, it requires no additional vent pipe.

The techniques for plumbing a half bath are similar to those used for a master bathroom. Refer to pages 100 to 107 for more detailed information.

In our half bath, the toilet and sink are close to the main stack for ease of installation, but are spaced far enough apart to meet minimum allowed distances between fixtures. Check your local Code for any restrictions in your area. Generally, there should be at least 15" from the center of the toilet drain to a side wall or fixture, and a minimum of 21" of space between the front edge of the toilet and the wall.

How to Plumb a Half Bath

1 Locate the main waste-vent stack in the wet wall, and remove the wall surface behind the planned location for the toilet and sink. Cut a 4½"-diameter hole for the toilet flange (centered 12" from the wall, for most toilets). Drill two ¾" holes through the sole plate for sink supply lines and one hole for the toilet supply line. Drill a 2" hole for the sink drain.

2 In the basement, cut away a section of the stack and insert two waste T-fittings. The top fitting should have a 3" side inlet for the toilet drain; the bottom fitting requires a 1½" reducing bushing for the sink drain. Install a toilet bend and 3" drain pipe for the toilet, and install a 1½" drain pipe with a sweep elbow for the sink.

3 Tap into water distribution pipes with ¾" × ½" reducing T-fittings, then run ½" copper supply pipes through the holes in the sole plate to the sink and toilet. Support all pipes at 4-ft. intervals with strapping attached to joists.

4 Attach drop ear elbows to the ends of the supply pipes, and anchor them to blocking installed between studs. Anchor the drain pipe to the blocking, then run a vertical vent pipe from the waste T-fitting up the wall to a point at least 6" above the highest fixture on the main stack. Then, route the vent pipe horizontally and splice it into the vent stack with a vent T.

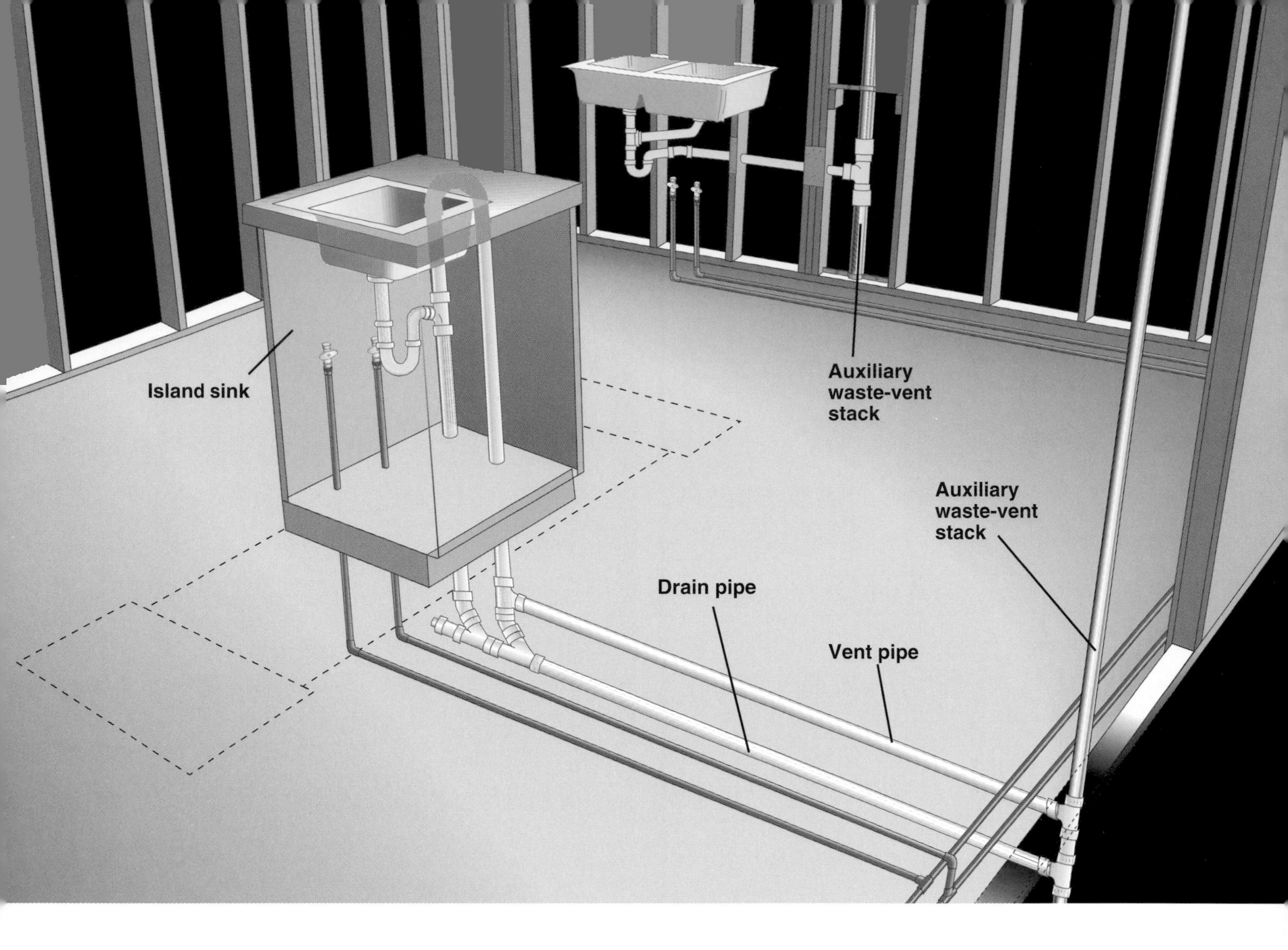

Plumbing a Kitchen

Plumbing a remodeled kitchen is a relatively easy job if your kitchen includes only a wall sink. If your project includes an island sink, however, the work becomes more complicated.

An island sink poses problems because there is no adjacent wall in which to run a vent pipe. For an island sink, you will need to use a special plumbing configuration known as a *loop vent*.

Each loop vent situation is different, and your configuration will depend on the location of existing waste-vent stacks, the direction of the floor joists, and the size and location of your sink base cabinet. Consult your local plumbing inspector for help in laying out the loop vent.

For our demonstration kitchen, we have divided the project into three phases:

- How to Install DWV Pipes for a Wall Sink (pages 118 to 120)
- How to Install DWV Pipes for an Island Sink (pages 121 to 125)
- How to Install New Supply Pipes (pages 126 to 127)

Our demonstration kitchen includes a double wall sink and an island sink. The 1½" drain for the wall sink connects to an existing 2" galvanized waste-vent stack; since the trap is within 3½ ft. of the stack, no vent pipe is required. The drain for the island sink uses a loop vent configuration connected to an auxiliary waste-vent stack in the basement.

Tips for Plumbing a Kitchen

Insulate exterior walls if you live in a region with freezing winter temperatures. Where possible, run water supply pipes through the floor or interior partition walls, rather than exterior walls.

Use existing waste-vent stacks to connect the new DWV pipes. In addition to a main waste-vent stack, most homes have one or more auxiliary waste-vent stacks in the kitchen that can be used to connect new DWV pipes.

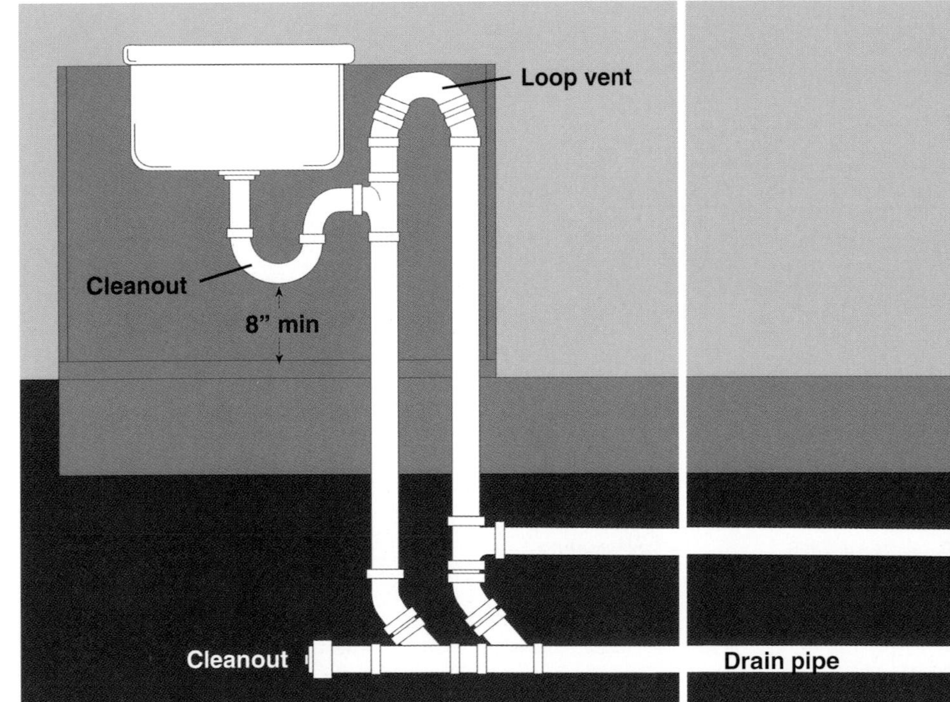

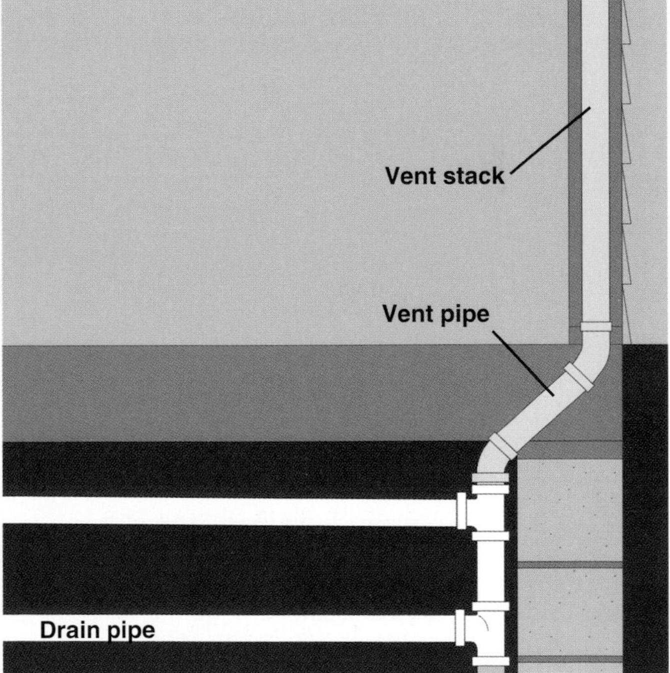

Loop vent makes it possible to vent a sink when there is no adjacent wall to house the vent pipe. The drain is vented with a loop of pipe that arches up against the countertop and away from the drain before dropping through the floor. The vent pipe then runs horizontally to an existing vent pipe. In our project, we have tied the island vent to a vent pipe extending up from a basement utility sink. NOTE: Loop vents are subject to local Code restrictions. Always consult your building inspector for guidelines on venting an island sink.

How to Install DWV Pipes for a Wall Sink

1 Determine the location of the sink drain by marking the position of the sink and base cabinet on the floor. Mark a point on the floor indicating the position of the sink drain opening. This point will serve as a reference for aligning the sink drain stub-out.

2 Mark a route for the new drain pipe through the studs behind the wall sink cabinet. The drain pipe should angle ¼" per foot down toward the waste-vent stack.

3 Use a right-angle drill and hole saw to bore holes for the drain pipe (page 96). On non-loadbearing studs, such as the cripple studs beneath a window, you can notch the studs with a reciprocating saw to simplify the installation of the drain pipe. If the studs are loadbearing, however, you must thread the run though the bored holes, using couplings to join short lengths of pipe as you create the run.

4 Measure, cut, and dry-fit a horizontal drain pipe to run from the waste-vent stack to the sink drain stub-out. Create the stub-out with a 45° elbow and 6" length of 1½" pipe. NOTE: If the sink trap in your installation will be more than 3½ ft. from the waste-vent pipe, you will need to install a waste T and run a vent pipe up the wall, connecting it to the vent stack at a point at least 6" above the lip of the sink.

5 Remove the neoprene sleeve from a banded coupling, then roll the lip back and measure the thickness of the separator ring.

6 Attach two lengths of 2" pipe, at least 4" long, to the top and bottom openings on a 2" × 2" × 1½" waste T. Hold the fitting alongside the waste-vent stack, then mark the stack for cutting, allowing space for the separator rings on the banded couplings.

(continued next page)

7 Use riser clamps and 2 × 4 blocking to support the waste-vent stack above and below the new drain pipe, then cut out the waste-vent stack along the marked lines, using a reciprocating saw and metal-cutting blade.

8 Slide banded couplings onto the cut ends of the waste-vent stack, and roll back the lips of the neoprene sleeves. Position the waste T assembly, then roll the sleeves into place over the plastic pipes.

9 Slide the metal bands into place over the neoprene sleeves, and tighten the clamps with a ratchet wrench or screwdriver.

10 Solvent-glue the drain pipe, beginning at the waste-vent stack. Use a 90° elbow and a short length of pipe to create a drain stub-out extending about 4" out from the wall.

How to Install DWV Pipes for an Island Sink

1 Position the base cabinet for the island sink, according to your kitchen plans. Mark the cabinet position on the floor with tape, then move the cabinet out of the way.

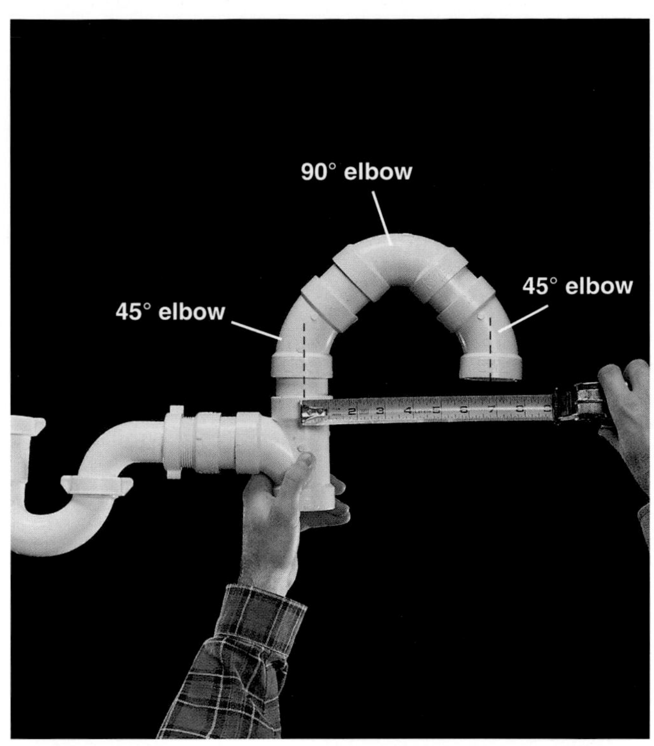

2 Create the beginning of the drain and loop vent by test-fitting a drain trap, waste T, two 45° elbows, and a 90° elbow, linking them with 2" lengths of pipe. Measure the width of the loop between the centerpoints of the fittings.

3 Draw a reference line perpendicular to the wall to use as a guide when positioning the drain pipes. A cardboard template of the sink can help you position the loop vent inside the outline of the cabinet.

4 Position the loop assembly on the floor, and use it as a guide for marking hole locations. Make sure to position the vent loop so the holes are not over joists.

(continued next page)

How to Install DWV Pipes for an Island Sink (continued)

5 Use a hole saw with a diameter slightly larger than the vent pipes to bore holes in the subfloor at the marked locations. Note the positions of the holes by carefully measuring from the edges of the taped cabinet outline; these measurements will make it easier to position matching holes in the floor of the base cabinet.

6 Reposition the base cabinet, and mark the floor of the cabinet where the drain and vent pipes will run. (Make sure to allow for the thickness of the cabinet sides when measuring.) Use the hole saw to bore holes in the floor of the cabinet, directly above the holes in the subfloor.

7 Measure, cut, and assemble the drain and loop vent assembly. Tape the top of the loop in place against a brace laid across the top of the cabinet, then extend the drain and vent pipes through the holes in the floor of the cabinet. The waste T should be about 18" above the floor, and the drain and vent pipes should extend about 2 ft. through the floor.

8 In the basement, establish a route from the island vent pipe to an existing vent pipe. (In our project, we are using the auxiliary waste-vent stack near a utility sink.) Hold a long length of pipe between the pipes, and mark for T-fittings. Cut off the plastic vent pipe at the mark, then dry-fit a waste T-fitting to the end of the pipe.

9 Hold a waste T against the vent stack, and mark the horizontal vent pipe at the correct length. Fit the horizontal pipe into the waste T, then tape the assembly in place against the vent stack. The vent pipe should angle ¼" per foot down toward the drain.

10 Fit a 3" length of pipe in the bottom opening on the T-fitting attached to the vent pipe, then mark both the vent pipe and the drain pipe for 45° elbows. Cut off the drain and vent pipes at the marks, then dry-fit the elbows onto the pipes.

11 Extend both the vent pipe and drain pipe by dry-fitting 3" lengths of pipe and Y-fittings to the elbows. Using a carpenter's level, make sure the horizontal drain pipe will slope toward the waste-vent at a pitch of ¼" per ft. Measure and cut a short length of pipe to fit between the Y-fittings.

(continued next page)

How to Install DWV Pipes for an Island Sink (continued)

12 Cut a horizontal drain pipe to reach from the vent Y-fitting to the auxiliary waste-vent stack. Attach a waste T to the end of the drain pipe, then position it against the drain stack, maintaining a downward slope of ¼" per ft. Mark the auxiliary stack for cutting above and below the fittings.

13 Cut out the auxiliary stack at the marks. Use the T-fittings and short lengths of pipe to assemble an insert piece to fit between the cutoff ends of the auxiliary stack. The insert assembly should be about ½" shorter than the removed section of stack.

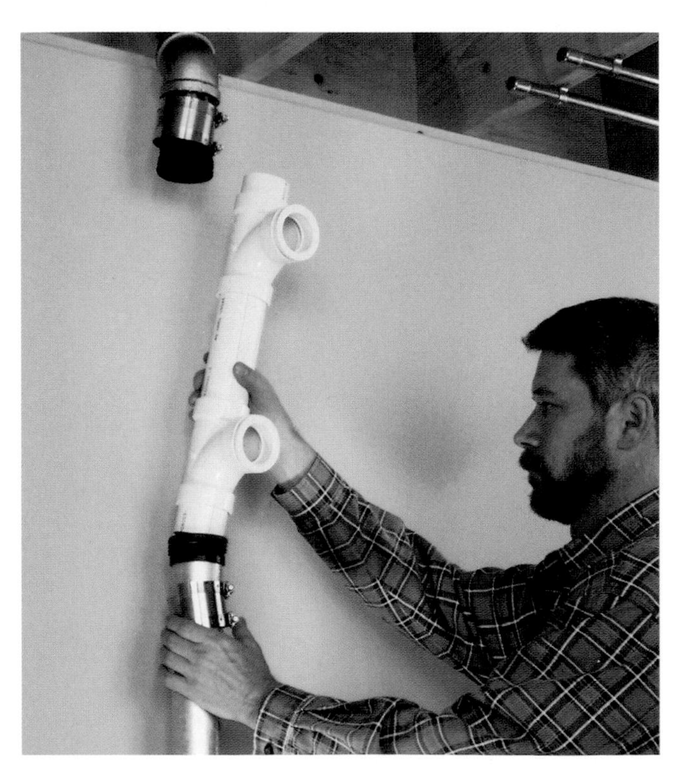

14 Slide banded couplings onto the cut ends of the auxiliary stack, then insert the plastic pipe assembly and loosely tighten the clamps.

15 At the open inlet on the drain pipe Y-fitting, insert a cleanout fitting.

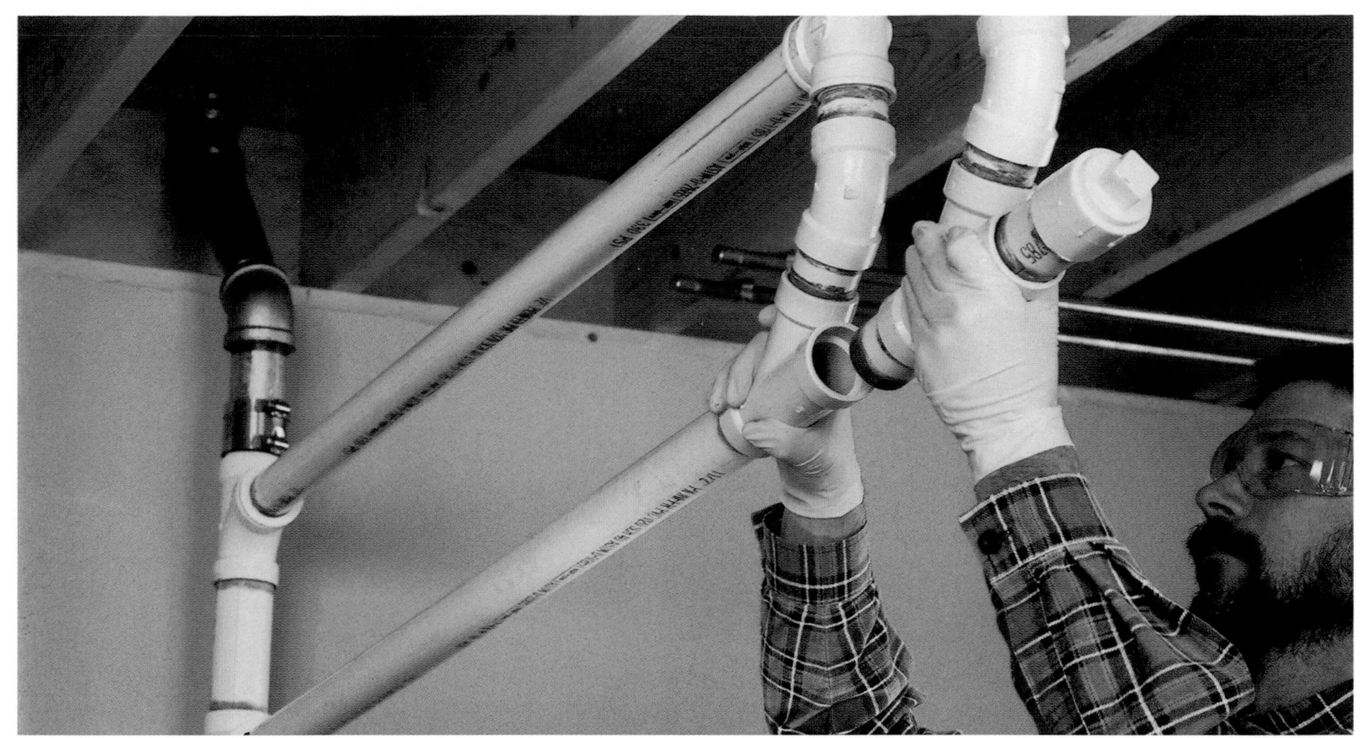

16 Solvent-glue all pipes and fittings found in the basement, beginning with the assembly inserted into the existing waste-vent stack, but do not glue the vertical drain and vent pipes running up into the cabinet. Tighten the banded couplings at the auxiliary stack. Support the horizontal pipes every 4 ft. with strapping nailed to the joists, then detach the vertical pipes extending up into the island cabinet. The final connection for the drain and vent loop will be completed as other phases of the kitchen remodeling project are finished.

17 After installing flooring and attaching cleats for the island base cabinet, cut away the flooring covering the holes for the drain and vent pipes.

18 Install the base cabinet, then insert the drain and vent pipes through the holes in the cabinet floor and solvent-glue the pieces together.

How to Install New Supply Pipes

1 Drill two 1"-diameter holes, spaced about 6" apart, through the floor of the island base cabinet and the underlying subfloor. Position the holes so they are not over floor joists. Drill similar holes in the floor of the base cabinet for the wall sink.

2 Turn off the water at the main shutoff, and drain the pipes. Cut out any old water supply pipes that obstruct new pipe runs, using a tubing cutter or hacksaw. In our project, we are removing the old pipe back to a point where it is convenient to begin the new branch lines.

3 Dry-fit T-fittings on each supply pipe (we used ¾" × ½" × ½" reducing T-fittings). Use elbows and lengths of copper pipe to begin the new branch lines running to the island sink and the wall sink. The parallel pipes should be routed so they are between 3" and 6" apart.

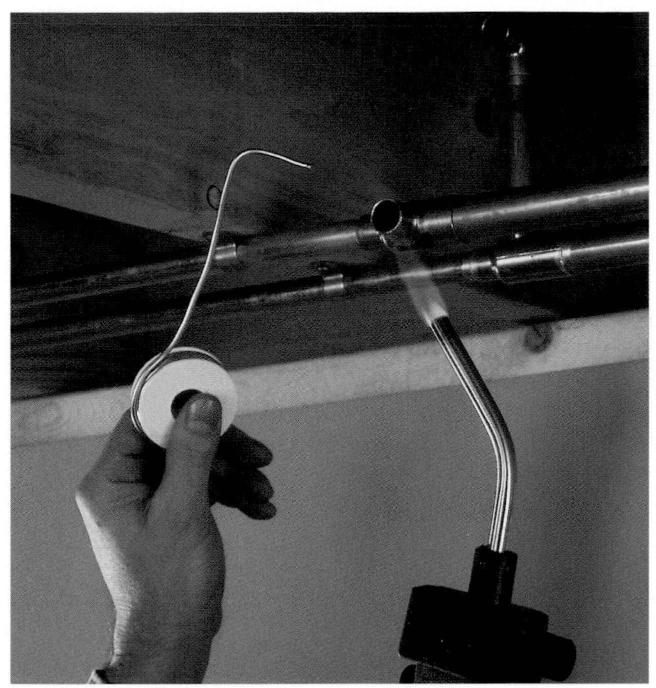

4 Solder the pipes and fittings together, beginning at the T-fittings. Support the horizontal pipe runs every 6 ft. with strapping attached to joists.

5 Extend the branch lines to points directly below the holes leading up into the base cabinets. Use elbows and lengths of pipe to form vertical risers extending at least 12" into the base cabinets. Use a small level to position the risers so they are plumb, then mark the pipe for cutting.

6 Fit the horizontal pipes and risers together, and solder them in place. Install blocking between joists, and anchor the risers to the blocking with pipe straps.

7 Solder male threaded adapters to the tops of the risers, then screw threaded shutoff valves onto the fittings.

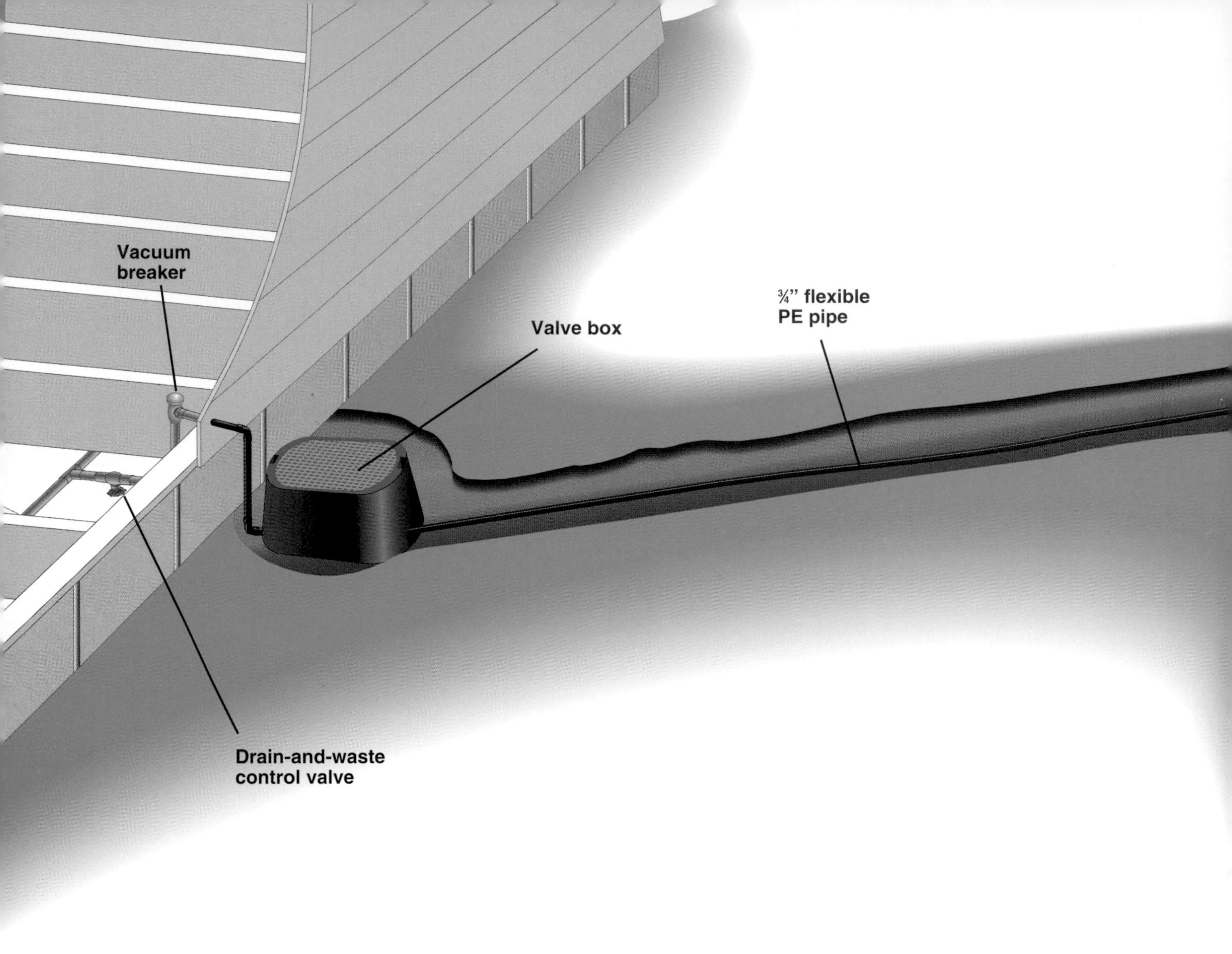

Installing Outdoor Plumbing

Flexible polyethylene (PE) pipe is used to extend cold-water plumbing lines to outdoor fixtures, such as a sink located in a shed or detached garage, a lawn sprinkler system, or garden spigot. In mild climates, outdoor plumbing can remain in service year-round, but in regions with a frost season, the outdoor supply pipes must be drained or blown empty with pressurized air to prevent the pipes from rupturing when the ground freezes in winter.

On the following pages you will see how to run supply pipes from the house to a utility sink in a detached garage. The utility sink drains into a rock-filled *dry well* installed in the yard. A dry well is designed to handle only "gray water" waste, such as the soapy rinse water created by washing tools or work clothes. Never use a dry well drain for septic materials, such as animal waste or food scraps. Never pour paints, solvent-based liquids, or solid materials into a sink that drains into a dry well. Such materials will quickly clog up your system and will eventually filter down into the groundwater supply.

Like an indoor sink, the garage utility sink has a vent pipe running up from the drain trap. This vent can extend through the roof (page 149), or it can be extended through the side wall of the garage and covered with a screen to keep birds and insects out.

Before digging a trench for an outdoor plumbing line, contact your local utility companies and ask them to mark the locations of underground gas, power, telephone, and water lines.

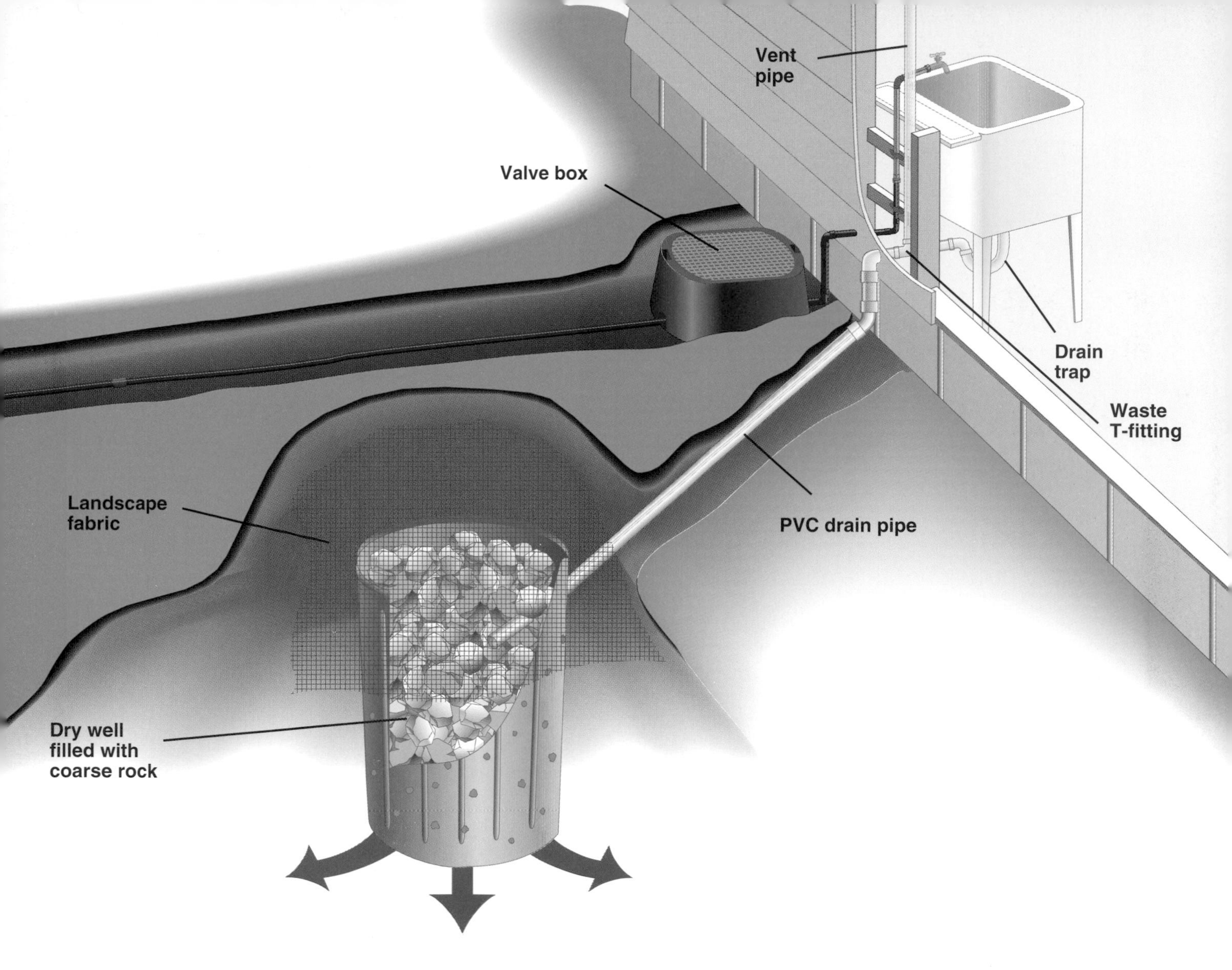

Underground lawn sprinkler systems can be installed using the same basic techniques used for plumbing an outdoor utility sink. Sprinkler systems vary from manufacturer to manufacturer, so make sure to follow the product recommendations when planning your system. See page 133.

How to Install Outdoor Plumbing for a Garage Sink

1 Plan a convenient route from a basement ¾" cold-water supply pipe to the outdoor sink location, then drill a 1"-diameter hole through the sill plate. Drill a similar hole where the pipe will enter the garage. On the ground outside, lay out the pipe run with spray paint or stakes.

2 Use a flat spade to remove sod for an 8"- to 12"-wide trench along the marked route from the house to the garage. Set the sod aside and keep it moist so it can be reused after the project is complete. Dig a trench that slopes slightly (⅛" per foot) toward the house and is at least 10" deep at its shallowest point. Use a long straight 2 × 4 and level to ensure that the trench has the correct slope.

3 Below the access hole in the rim joist, dig a small pit and install a plastic valve box so the top is flush with the ground. Lay a thick layer of gravel in the bottom of the box. Dig a similar pit and install a second valve box at the opposite end of the trench, where the water line will enter the garage.

4 Run ¾" PE pipe along the bottom of the trench from the house to the utility sink location. Use insert couplings and stainless steel clamps when it is necessary to join two lengths of pipe.

Tip: To run pipe under a sidewalk, attach a length of rigid PVC pipe to a garden hose with a pipe-to-hose adapter. Cap the end of the pipe, and drill a ⅛" hole in the center of the cap. Turn on the water, and use the high-pressure stream to bore a tunnel.

5 At each end of the trench, extend the pipe through the valve box and up the foundation wall, using a barbed elbow fitting to make the 90° bend. Install a barbed T-fitting with a threaded outlet in the valve box, so the threaded portion of the fitting faces down. Insert a male threaded plug in the bottom outlet of the T-fitting.

6 Use barbed elbow fittings to extend the pipe into the basement and garage, then use pipe straps and masonry screws to anchor the PE pipe to the foundation.

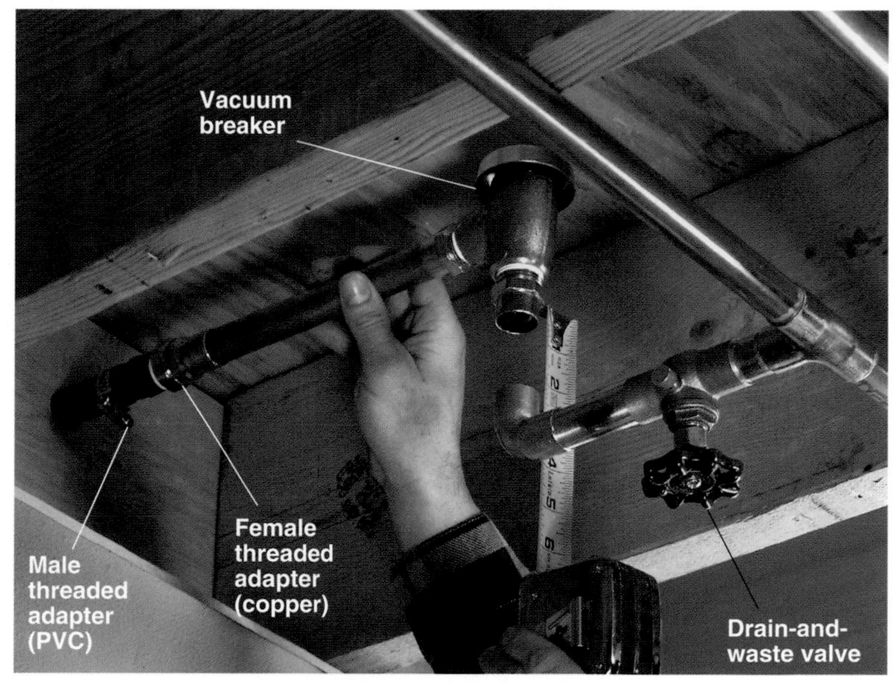

7 Inside the house, make the transition between the PE pipe and the copper cold-water supply pipe, using a threaded male PVC adapter, a female threaded copper adapter, a vacuum breaker, a drain-and-waste valve, and a copper T-fitting, as shown. The drain-and-waste valve includes a threaded cap, which can be removed to blow water from the lines when you are winterizing the system.

(continued next page)

How to Install Outdoor Plumbing for a Garage Sink (continued)

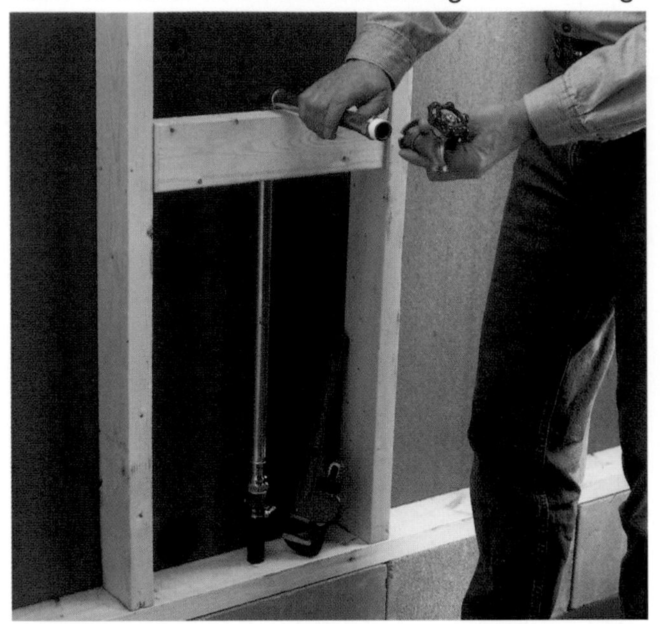

8 In the garage, attach a male threaded PVC adapter to the end of the PE pipe, then use a copper female threaded adapter, elbow, and male threaded adapter to extend a copper riser up to a brass hose bib. After completing the supply pipe installation, fill in the trench, tamping the soil firmly. Install the utility sink, complete with 1½" drain trap and waste T-fitting (page 129). Bore a 2" hole in the wall where the sink drain will exit the garage.

9 At least 6 ft. from the garage, dig a pit about 2 ft. in diameter and 3 ft. deep. Punch holes in the sides and bottom of an old trash can, and cut a 2" hole in the side of the can, about 4" from the top edge. Insert the can into the pit; the top edge should be about 6" below ground level. Run 1½" PVC drain pipe from the utility sink to the dry well. Fill the dry well with coarse rock, drape landscape fabric over it, then cover the trench and well with soil and reinstall the sod. Extend a vent pipe up from the waste T through the roof or side wall of the garage (page 149).

How to Winterize Outdoor Plumbing Pipes in Cold Climates

Close the drain-and-waste valve for the outdoor supply pipe, then remove the cap on the drain nipple. With the hose bib on the outdoor sink open, attach an air compressor to the valve nipple, then blow water from the system using no more than 50 psi (pounds per square inch) of air pressure. Remove the plugs from the T-fittings in each valve box, and store them for the winter.

Components of an Underground Sprinkler System

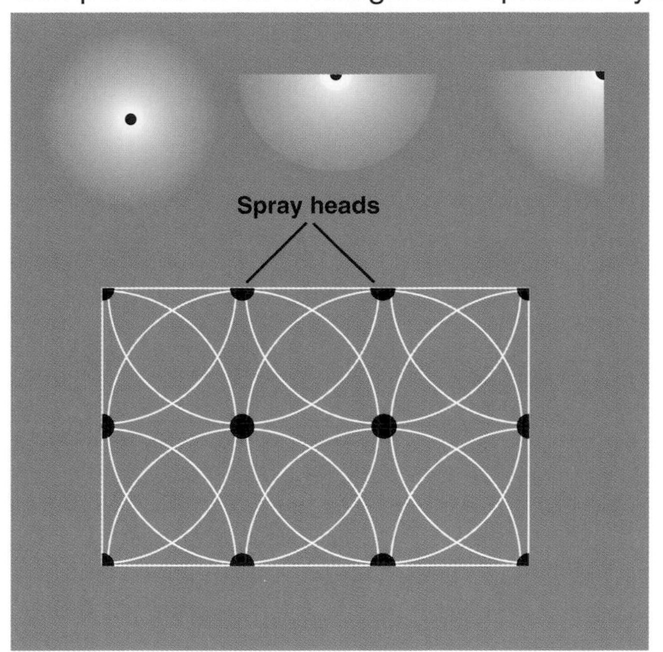

System layout is crucial to ensure proper irrigation of all parts of your yard. Spray heads are available in full-circle, semicircle, or quarter-circle patterns to cover the entire space. In most cases, you will divide your landscape into several individual zones, each controlled by its own valve. With a timer (photo below), you can program precise start and stop times for each irrigation zone.

Photo courtesy of Hunter Industries

Valve manifold is a group of valves used to control the various sprinkler zones. Some models are installed below ground in a valve box, while others extend above the ground. When a timer is used, each control valve in the manifold is wired separately into the timer.

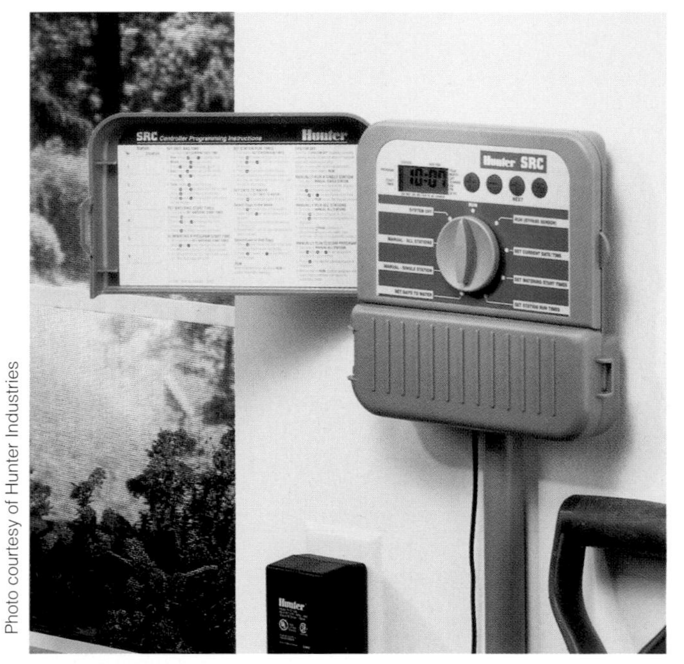

Photo courtesy of Hunter Industries

Sprinkler timers can be programmed to provide automatic control of all zones in an underground sprinkler system. Deluxe models control up to 16 different zones and have rain sensors that shut off the system when irrigation is not needed.

Photo courtesy of Hunter Industries

Sprinkler heads come in many styles to provide a variety of spray patterns. Flexible underground tubing, sometimes called *funny pipe*, links the spray head to saddle fittings on the underground water supply pipes.

Leave old plumbing pipes in place, if possible. To save time, professional plumbing contractors remove old plumbing pipes only when they interfere with the routing of the new plumbing lines.

Replacing Old Plumbing

Plumbing pipes, like all building materials, eventually wear out and have to be replaced. If you find yourself repairing leaky, corroded pipes every few months, it may be time to consider replacing the old system entirely—and soon. A corroded water pipe that bursts while you are away can cost you many thousands of dollars in damage to wall surfaces, framing members, and furnishings.

Identifying the materials used in your plumbing system can also tell you if replacement is advised. If you have galvanized steel pipes, for example, it is a good bet that they will need to be replaced in the near future. Most galvanized steel pipes were installed before 1960, and since steel pipes have a maximum life expectancy of 30 to 35 years, such a system is probably living on borrowed time. On the other hand, if your system includes copper supply pipes and plastic drain pipes, you can relax; these materials were likely installed within the last 30 years, and they are considerably more durable than steel, provided they were installed correctly.

Unless you live in a rambler with an exposed basement ceiling, replacing old plumbing nearly always involves some demolition and carpentry work. Even in the best scenario, you probably will find it necessary to open walls and floors in order to run new pipes. For this reason, replacing old plumbing is often done at the same time as a kitchen or bathroom remodeling project, when wall and floor surfaces have to be removed and replaced.

A plumbing renovation project is subject to the same Code regulations as a new installation. Always work in conjunction with your local inspector (pages 86 to 91) when replacing old plumbing.

This section shows:

- Evaluating Your Plumbing (pages 136 to 137)
- Step-by-step Overview (pages 138 to 139)
- Planning Pipe Routes (pages 140 to 143)
- Replacing a Main Waste-Vent Stack (pages 144 to 149)
- Replacing Branch Drains & Vent Pipes (pages 150 to 153)
- Replacing a Toilet Drain (pages 154 to 155)
- Replacing Supply Pipes (pages 156 to 157)

Replacement Options

Partial replacement involves replacing only those sections of your plumbing system that are currently causing problems. This is a quick, less expensive option than a complete renovation, but it is only a temporary solution. Old plumbing will continue to fail until you replace the entire system.

Complete replacement of all plumbing lines is an ambitious job, but doing this work yourself can save you thousands of dollars. To minimize the inconvenience, you can do this work in phases, replacing one branch of the plumbing system at a time.

Evaluating Your Plumbing

By the time you spot the telltale evidence of a leaky drain pipe or water supply pipe, the damage to the walls and ceilings of your home can be considerable. The tips on the following pages show early warning signals that indicate your plumbing system is beginning to fail.

Proper evaluation of your plumbing helps you identify old, suspect materials and anticipate problems. It also can save you money and aggravation. Replacing an old plumbing system at your convenience before it reaches the disaster stage is preferable to hiring a plumbing contractor to bail you out of an emergency situation.

Remember that the network of pipes running through the walls of your home is only one part of the larger system. You should also evaluate the main water supply and sewer pipes that connect your home to the city utility system and make sure they are adequate before you replace your plumbing.

Fixture Units	Minimum Gallons per Minute (GPM)
10	8
15	11
20	14
25	17
30	20

Minimum recommended water capacity is based on total demand on the system, as measured by fixture units, a standard of measurement assigned by the Plumbing Code. First, add up the total units of all the fixtures in your plumbing system (page 88). Then, perform the water supply capacity test described below. Finally, compare your water capacity with the recommended minimums listed above. If the capacity falls below that recommended in the table above, then the main water supply pipe running from the city water main to your home is inadequate and should be replaced with a larger pipe by a licensed contractor.

How to Determine Your Water Supply Capacity

1 Shut off the water at the valve on your water meter, then disconnect the pipe on the house side of the meter. Construct a test spout using a 2" PVC elbow and two 6" lengths of 2" PVC pipe, then place the spout on the exposed outlet on the water meter. Place a large watertight tub under the spout to collect water.

2 Open the main supply valve and let the water run into the container for 30 seconds. Shut off the water, then measure the amount of water in the container and multiply this figure by 2. This number represents the gallons-per-minute rate of your main water supply. Compare this measurement with the recommended capacity in the table above.

Symptoms of Bad Plumbing

Rust stains on the surfaces of toilet bowls and sinks may indicate severe corrosion inside iron supply pipes. This symptom generally means your water supply system is likely to fail in the near future. NOTE: Rust stains can also be caused by a water heater problem or by a water supply with a high mineral content. Check for these problems before assuming your pipes are bad.

Low water pressure at fixtures suggests that the supply pipes either are badly clogged with rust and mineral deposits, or are undersized. To measure water pressure, plug the fixture drain and open the faucets for 30 seconds. Measure the amount of water and multiply by 2; this figure is the gallons-per-minute rating. Vanity faucets should supply 1¾ gpm; bathtub faucets, 6 gpm; kitchen sink faucets, 4½ gpm.

Slow drains throughout the house may indicate that DWV pipes are badly clogged with rust and mineral deposits. When a fixture faucet is opened fully with the drains unstopped, water should not collect in tubs and basins. NOTE: Slow drains may also be the result of inadequate venting. Check for this problem before assuming the drain pipes are bad.

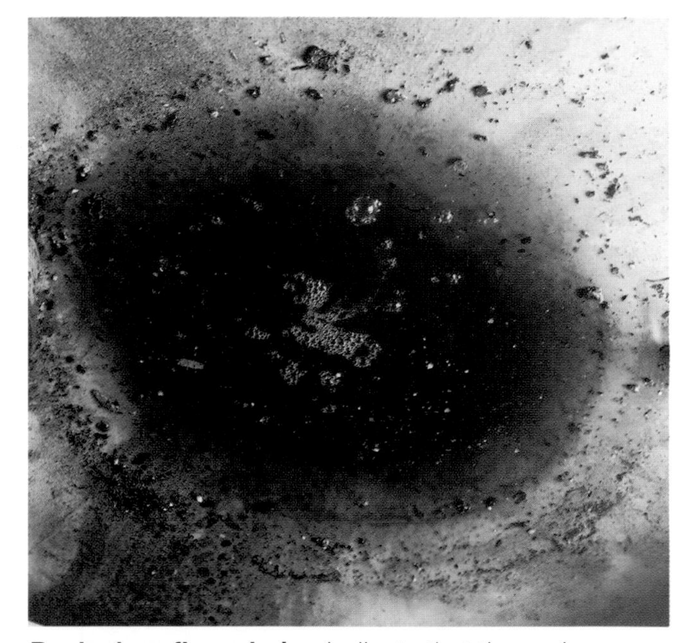

Backed-up floor drains indicate that the main sewer service to the street is clogged. If you have this problem regularly, have the main sewer lines evaluated by a plumbing contractor before you replace your house plumbing. The contractor will be able to determine if your sewer problem is a temporary clog or a more serious problem that requires major work.

Replacing Old Plumbing: A Step-by-step Overview

The overview sequence shown here represents the basic steps you will need to follow when replacing DWV and water supply pipes. On the following pages, you will see these steps demonstrated in complete detail, as we replace all the water supply pipes and drain pipes for a bathroom, including the main waste-vent stack running from basement to roofline.

Remember that no two plumbing jobs are ever alike, and your own project will probably differ from the demonstration projects shown in this section. Always work in conjunction with your local plumbing inspector, and organize your work around a detailed plumbing plan that shows the particulars of your project. Review pages 74 to 93 of this book before starting work.

1 Plan the routes for the new plumbing pipes. Creating efficient pathways for new pipes is crucial to a smooth installation. In some cases this requires removing wall or floor surfaces. Or, you can frame a false wall, called a *chase* (page 140), to create space for running the new pipes.

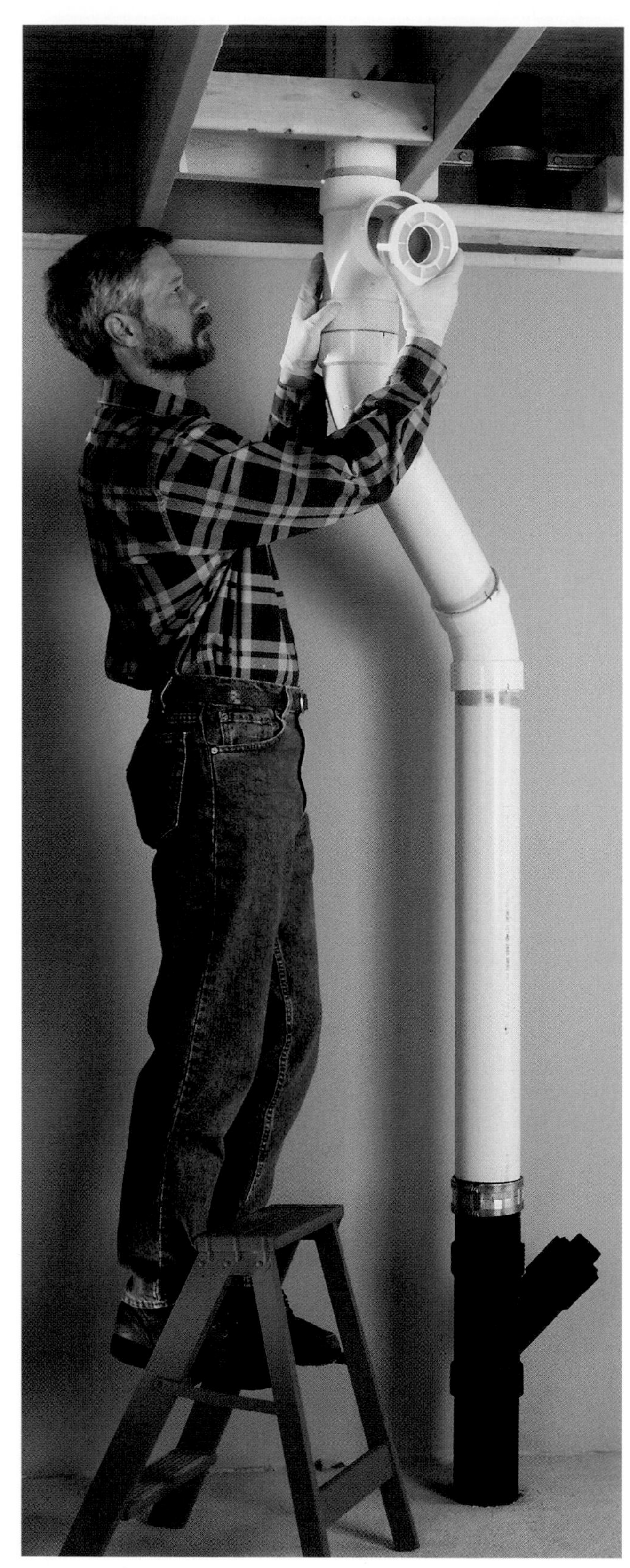

2 Remove sections of the old waste-vent stack, as needed, then install a new main waste-vent stack running from the main drain in the basement to the roof. Include the fittings necessary to connect branch drains and vent lines to the stack.

3 Install new branch drains from the waste-vent stack to the stub-outs for the individual fixtures. If the fixture locations have not changed, you may need to remove sections of the old drain pipes in order to run the new pipes.

4 Remove the old toilet bend and replace it with a new bend running to the new waste-vent stack. This task usually requires that you remove areas of flooring. Framing work may also be required to create a path for the toilet drain.

5 Replace the vent pipes running from the fixtures up to the attic, then connect them to the new waste-vent stack.

6 Install new copper supply lines running from the water meter to all fixture locations. Test the DWV and water supply pipes and have your work inspected before closing up walls and installing the fixtures.

Planning Pipe Routes

Build a framed chase. A chase is a false wall created to provide space for new plumbing pipes. It is especially effective for installing a new main waste-vent stack. On a two-story house, chases can be stacked one over the other on each floor in order to run plumbing from the basement to the attic. Once plumbing is completed and inspected, the chase is covered with wallboard and finished to match the room.

The first, and perhaps most important, step when replacing old plumbing is to decide how and where to run the new pipes. Since the stud cavities and joist spaces are often covered with finished wall surfaces, finding routes for running new pipes can be challenging.

When planning pipe routes, choose straight, easy pathways whenever possible. Rather than running water supply pipes around wall corners and through studs, for example, it may be easiest to run them straight up wall cavities from the basement. Instead of running a bathtub drain across floor joists, run it straight down into the basement, where the branch drain can be easily extended underneath the joists to the main waste-vent stack.

In some situations, it is most practical to route the new pipes in wall and floor cavities that already hold plumbing pipes, since these spaces often are framed to provide long, unobstructed runs. A detailed map of your plumbing system can be very helpful when planning routes for new plumbing pipes (pages 76 to 81).

To maximize their profits, plumbing contractors generally try to avoid opening walls or changing wall framing when installing new plumbing. But the do-it-yourselfer does not have these limitations. Faced with the difficulty of running pipes through enclosed spaces, you may find it easiest to remove wall surfaces or to create a newly framed space for running new pipes.

On these pages, you will see some common methods used to create pathways for replacing old pipes with new plumbing.

Tips for Planning Pipe Routes

Use existing access panels to disconnect fixtures and remove old pipes. Plan the location of new fixtures and pipe runs to make use of existing access panels, minimizing the amount of demolition and repair work you will need to do.

Convert a laundry chute into a channel for running new plumbing pipes. The door of the chute can be used to provide access to control valves, or it can be removed and covered with wall materials, then finished to match the surrounding wall.

Run pipes inside a closet. If they are unobtrusive, pipes can be left exposed at the back of the closet. Or, you can frame a chase to hide the pipes after the installation is complete.

Remove false ceiling panels to route new plumbing pipes in joist cavities. Or, you can route pipes across a standard plaster or wallboard ceiling, then construct a false ceiling to cover the installation, provided there is adequate height. Most Building Codes require a minimum of 7 ft. from floor to finished ceiling.

(continued next page)

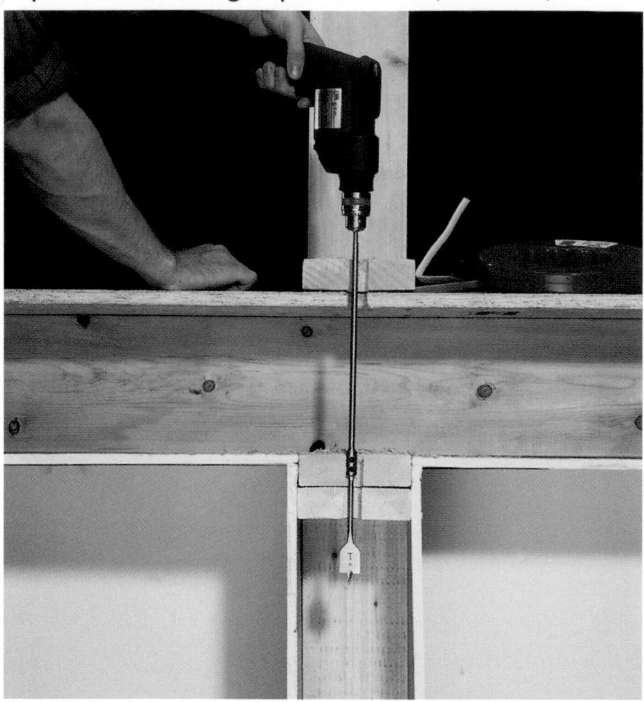

Use a drill extension and spade bit or hole saw to drill through wall plates from unfinished attic or basement spaces above or below the wall.

Look for "wet walls." Walls that hold old plumbing pipes can be good choices for running long vertical runs of new pipe. These spaces usually are open, without obstacles such as fireblocks and insulation.

Probe wall and floor cavities with a long piece of plastic pipe to ensure that a clear pathway exists for running new pipe (left). Once you have established a route using the narrow pipe, you can use the pipe as a guide when running larger drain pipes up into the wall (right).

Remove flooring when necessary. Because replacing toilet and bathtub drains usually requires that you remove sections of floor, a full plumbing replacement job is often done in conjunction with a complete bathroom remodeling project.

Remove wall surfaces when access from above or below the wall is not possible. This demolition work can range from cutting narrow channels in plaster or wallboard to removing the entire wall surface. Remove wall surfaces back to the centers of adjoining studs; the exposed studs provide a nailing surface for attaching new wall materials once the plumbing project is completed.

Create a detailed map showing the planned route for your new plumbing pipes. Such a map can help you get your plans approved by the inspector, and it makes work much simpler. If you have already mapped your existing plumbing system (pages 76 to 81), those drawings can be used to plan new pipe routes.

A new main waste-vent stack is best installed near the location of the old stack. In this way, the new stack can be connected to the basement floor cleanout fitting used by the old cast-iron stack.

Replacing a Main Waste-Vent Stack

Although a main waste-vent stack rarely rusts through entirely, it can be nearly impossible to join new branch drains and vents to an old cast-iron stack. For this reason, plumbing contractors sometimes recommend replacing the iron stack with plastic pipe during a plumbing renovation project.

Be aware that replacing a main waste-vent stack is not an easy job. You will be cutting away heavy sections of cast iron, so working with a helper is essential. Before beginning work, make sure you have a complete plan for your plumbing system and have designed a stack that includes all the fittings you will need to connect branch drains and vent pipes. While work is in progress, none of your plumbing fixtures will be usable. To speed up the project and minimize inconvenience, do as much of the demolition and preliminary construction work as you can before starting work on the stack.

Because main waste-vent stacks may be as large as 4" in diameter, running a new stack through existing walls can be troublesome. To solve this problem, our project employs a common solution: framing a chase in the corner of a room to provide the necessary space for running the new stack from the basement to the attic. When the installation is complete, the chase will be finished with wallboard to match the room.

How to Replace a Main Waste-Vent Stack

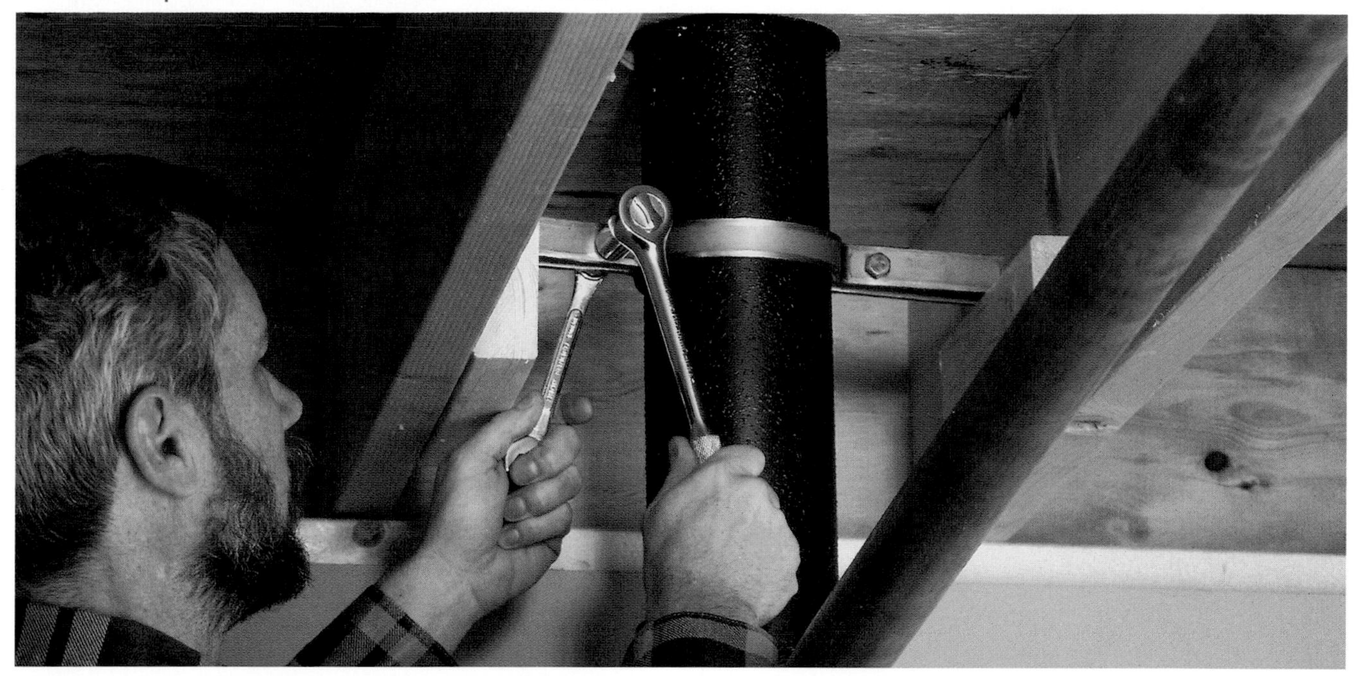

1 Secure the cast-iron waste-vent stack near the ceiling of your basement, using a riser clamp installed between the floor joists. Use wood blocks attached to the joists with 3" wallboard screws to support the clamp. Also clamp the stack in the attic, at the point where the stack passes down into the wall cavity. WARNING: A cast-iron stack running from basement to attic can weigh several hundred pounds. Never cut into a cast-iron stack before securing it with riser clamps above the cut.

2 Use a cast iron snap cutter to sever the stack near the floor of the basement, about 8" above the cleanout, and near the ceiling, flush with the bottom of the joists. Have a helper hold the stack while you are cutting out the section. NOTE: After cutting into the main waste-vent stack, plug the open end of the pipe with a cloth to prevent sewer gases from rising into your home.

3 Nail blocking against the bottom of the joists across the severed stack. Then, cut a 6"-diameter hole in the basement ceiling where the new waste-vent stack will run, using a reciprocating saw. Suspend a plumb bob at the centerpoint of the opening as a guide for aligning the new stack.

(continued next page)

4 Attach a 5-ft. segment of PVC plastic pipe the same diameter as the old waste-vent stack to the exposed end of the cast-iron cleanout fitting, using a banded coupling with neoprene sleeve.

5 Dry-fit 45° elbows and straight lengths of plastic pipe to offset the new stack, lining it up with the plumb bob centered on the ceiling opening.

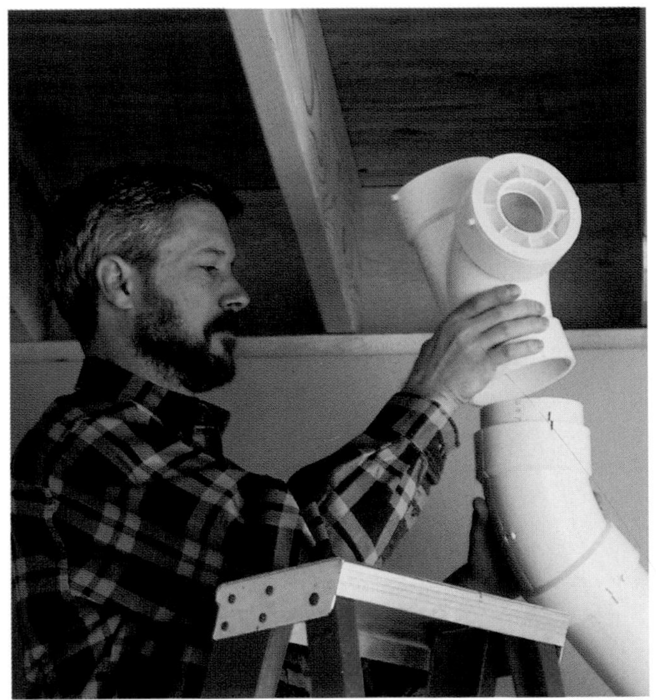

6 Dry-fit a waste T-fitting on the stack, with the inlets necessary for any branch drains that will be connected in the basement. Make sure the fitting is positioned at a height that will allow the branch drains to have the correct ¼" per foot downward slope toward the stack.

7 Determine the length for the next piece of waste-vent pipe by measuring from the basement T-fitting to the next planned fitting in the vertical run. In our project, we will be installing a T-fitting between floor joists, where the toilet drain will be connected.

8 Cut a PVC plastic pipe to length, raise it into the opening, and dry-fit it to the T-fitting. NOTE: For very long pipe runs, you may need to construct this vertical run by solvent-gluing two or more segments of pipe together with couplings.

9 Check the length of the stack, then solvent-glue all fittings together. Support the new stack with a riser clamp resting on blocks attached between basement ceiling joists.

10 Attach the next waste T-fitting to the stack. In our demonstration project, the waste T lies between floor joists and will be used to connect the toilet drain. Make sure the waste T is positioned at a height which will allow for the correct ⅛" per foot downward slope for the toilet drain.

11 Add additional lengths of pipe, with waste T-fittings installed where other fixtures will drain into the stack. In our example, a waste T with a 1½" bushing insert is installed where the vanity sink drain will be attached to the stack. Make sure the T-fittings are positioned to allow for the correct downward pitch of the branch drains.

(continued next page)

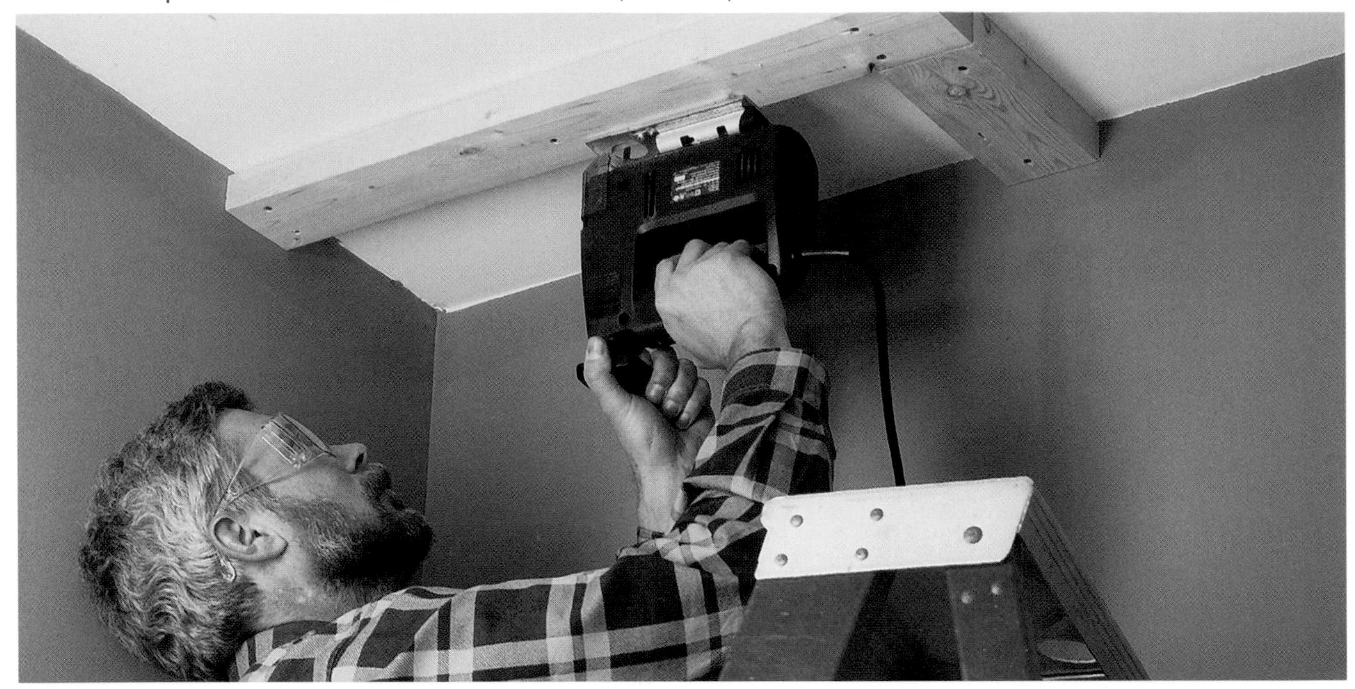

12 Cut a hole in the ceiling where the waste-vent stack will extend into the attic, then measure, cut, and solvent-glue the next length of pipe in place. The pipe should extend at least 1 ft. up into the attic.

13 Remove the roof flashing from around the old waste-vent stack. You may need to remove shingles in order to accomplish this. NOTE: Always use caution when working on a roof. If you are unsure of your ability to do this work, hire a roof repair specialist to remove the old flashing and install new flashing around the new vent pipe.

14 In the attic, remove old vent pipes, where necessary, then sever the cast-iron soil stack with a cast iron cutter and lower the stack down from the roof opening with the aid of a helper. Support the old stack with a riser clamp installed between joists.

15 Solvent-glue a vent T with a 1½" bushing in the side inlet to the top of the new waste-vent stack. The side inlet should point toward the nearest auxiliary vent pipe extending up from below.

16 Finish the waste-vent stack installation by using 45° elbows and straight lengths of pipe to extend the stack through the same roof opening used by the old vent stack. The new stack should extend at least 1 ft. through the roof, but no more than 2 ft.

How to Flash a Waste-Vent Stack

1 Loosen the shingles directly above the new vent stack, and remove any nails, using a flat pry bar. When installed, the metal vent flashing will lie flat on the shingles surrounding the vent pipe. Apply roofing cement to the underside of the flashing.

2 Slide the flashing over the vent pipe, and carefully tuck the base of the flashing up under the shingle. Press the flange firmly against the roof deck to spread the roofing cement, then anchor it with rubber gasket flashing nails. Reattach loose shingles as necessary.

Remove old pipes only where they obstruct the planned route for the new pipes. You will probably need to remove drain and water supply pipes at each fixture location, but the remaining pipes usually can be left in place. A reciprocating saw with metal-cutting blade works well for this job.

Replacing Branch Drains & Vent Pipes

In our demonstration project, we are replacing branch drains for a bathtub and vanity sink. The tub drain will run down into the basement before connecting to the main waste-vent stack, while the vanity drain will run horizontally to connect directly to the stack.

A vent pipe for the bathtub runs up into the attic, where it will join the main waste-vent stack. The vanity sink, however, requires no secondary vent pipe, since its location falls within the critical distance (page 91) of the new waste-vent stack.

How to Replace Branch Drains

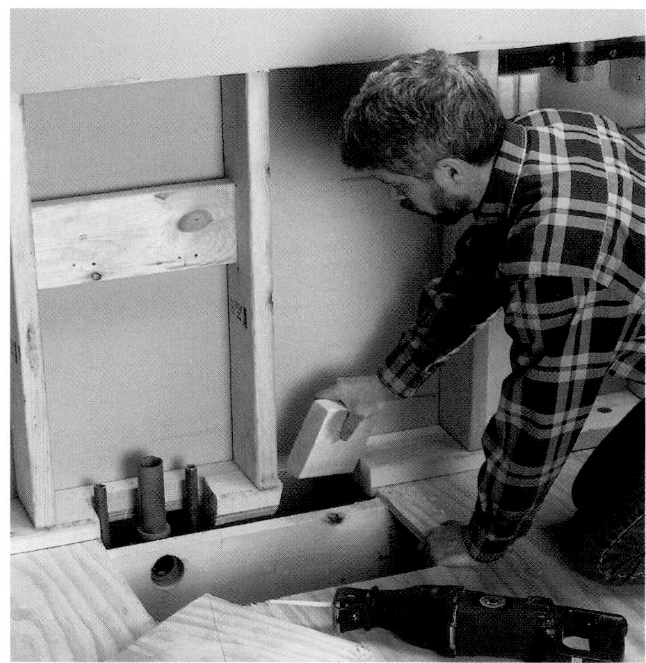

1 Establish a route for vertical drain pipes running through wall cavities down into the basement. For our project, we are cutting away a section of the wall sole plate in order to run a 1½" bathtub drain pipe from the basement up to the bathroom.

2 From the basement, cut a hole in the bottom of the wall, below the opening cut (step 1). Measure, cut, and insert a length of vertical drain pipe up into the wall to the bathroom. A length of flexible CPVC pipe can be useful for guiding the drain pipe up into the wall. For very long pipe runs, you may need to join two or more lengths of pipe with couplings as you insert the run.

3 Secure the vertical drain pipe with a riser clamp supported on 2 × 4 blocks nailed between joists. Take care not to overtighten the clamps.

4 Install a horizontal pipe from the waste T-fitting on the waste-vent stack to the vertical drain pipe. Maintain a downward slope toward the stack of ¼" per foot, and use a Y-fitting with 45° elbow to form a cleanout where the horizontal and vertical drain pipes meet.

5 Solvent-glue a waste T-fitting to the top of the vertical drain pipe. For a bathtub drain, as shown here, the T-fitting must be well below floor level to allow for the bathtub drain trap. You may need to notch or cut a hole in floor joists to connect the drain trap to the waste T (page 96).

6 From the attic, cut a hole into the top of the bathroom wet wall, directly above the bathtub drain pipe. Run a 1½" vent pipe down to the bathtub location, and solvent-glue it to the waste T. Make sure the pipe extends at least 1 ft. into the attic.

(continued next page)

7 Remove wall surfaces as necessary to provide access for running horizontal drain pipes from fixtures to the new waste-vent stack. In our project, we are running 1½" drain pipe from a vanity sink to the stack. Mark the drain route on the exposed studs, maintaining a ¼" per foot downward slope toward the stack. Use a reciprocating saw or jig saw to notch out the studs (page 96).

8 Secure the old drain and vent pipes with riser clamps supported by blocking attached between the studs.

9 Remove the old drain and water supply pipes, where necessary, to provide space for running the new drain pipes.

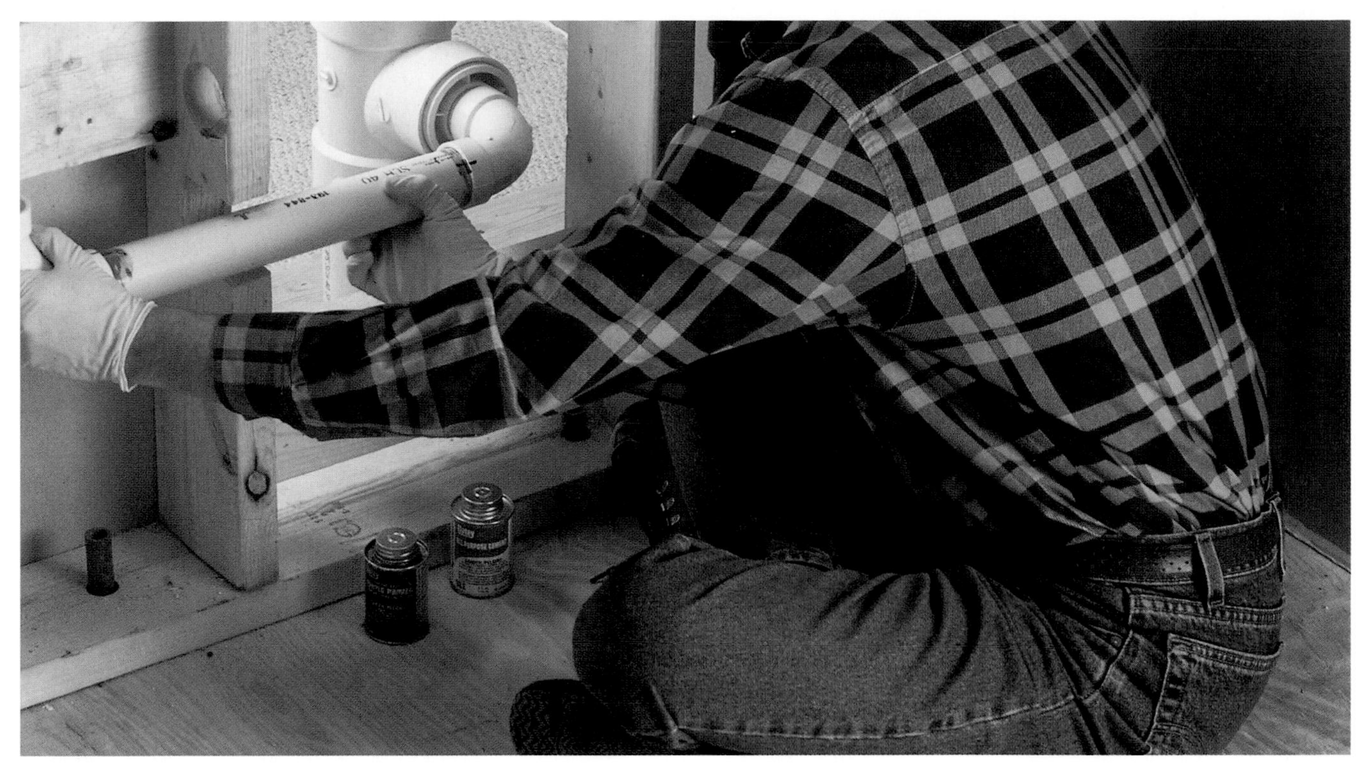

10 Using a sweep elbow and straight length of pipe, assemble a drain pipe to run from the drain stub-out location to the waste T-fitting on the new waste-vent stack. Use a 90° elbow and a short length of pipe to create a stub-out extending at least 2" out from the wall. Secure the stub-out to a ¾" backer board attached between studs.

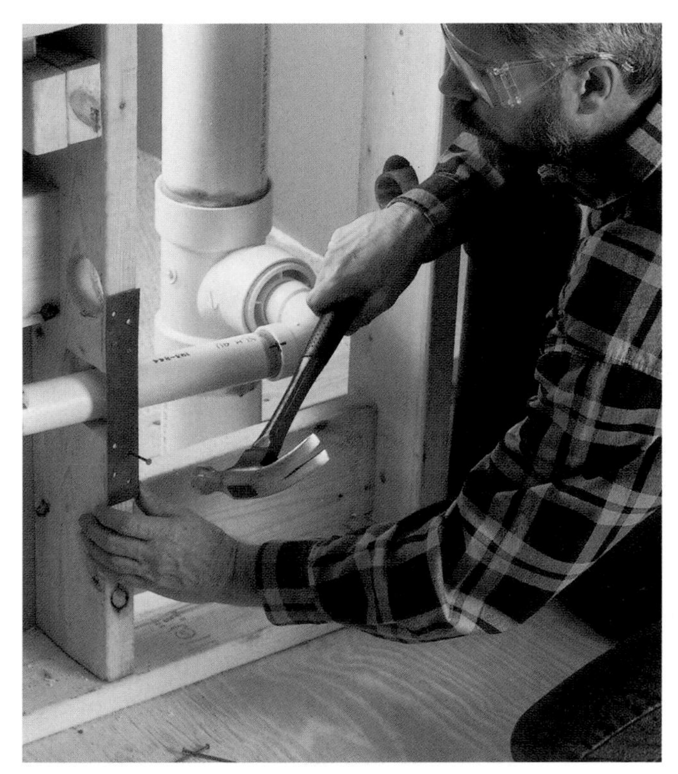

11 Protect the drain pipes by attaching metal protector plates over the notches in the studs. Protector plates prevent drain pipes from being punctured when wall surfaces are replaced.

12 In the attic, use a vent elbow and straight length of pipe to connect the vertical vent pipe from the tub to the new waste-vent stack.

Replacing a Toilet Drain

Replacing a toilet drain usually requires that you remove flooring and wall surface to gain access to the pipes.

Replacing a toilet drain is sometimes a troublesome task, mostly because the cramped space makes it difficult to route the large, 3" or 4" pipe. You likely will need to remove flooring around the toilet and wall surface behind the toilet.

Replacing a toilet drain may require framing work, as well, if you find it necessary to cut into joists in order to route the new pipes. When possible, plan your project to avoid changes to the framing members.

How to Replace a Toilet Drain

1 Remove the toilet, then unscrew the toilet flange from the floor and remove it from the drain pipe. NOTE: If the existing toilet flange is cast iron or bronze, it may be joined to the toilet bend with poured lead or solder; in this case, it is easiest to break up the flange with a masonry hammer (make sure to wear eye protection) and remove it in pieces.

2 Cut away the flooring around the toilet drain along the center of the floor joists, using a circular saw with the blade set to a depth ⅛" more than the thickness of the subfloor. The exposed joist will serve as a nailing surface when the subfloor is replaced.

3 Cut away the old toilet bend as close as possible to the old waste-vent stack, using a reciprocating saw with metal-cutting blade, or a cast iron cutter.

4 If a joist obstructs the route to the new waste-vent stack, cut away a section of the floor joist. Install double headers and metal joist hangers to support the ends of the cut joist.

5 Create a new toilet drain running to the new waste-vent stack, using a toilet bend and a straight length of pipe. Position the drain so there will be at least 15" of space between the center of the bowl and side wall surfaces when the toilet is installed. Make sure the drain slopes at least ⅛" per foot toward the stack, then support the pipe with plastic pipe strapping attached to framing members. Insert a 6" length of pipe in the top inlet of the closet bend; once the new drain pipes have been tested, this pipe will be cut off with a handsaw and fitted with a toilet flange.

6 Cut a piece of exterior-grade plywood to fit the cutout floor area, and use a jig saw to cut an opening for the toilet drain stub-out. Position the plywood, and attach it to joists and blocking with 2" screws.

Replacing Supply Pipes

Support copper supply pipes every 6 ft. along vertical runs and 10 ft. along horizontal runs. Always use copper or plastic support materials with copper; never use steel straps, which can interact with copper and cause corrosion.

When replacing old galvanized water supply pipes, we recommend that you use type-M rigid copper. Use ¾" pipe for the main distribution pipes and ½" pipes for the branch lines running to individual fixtures.

For convenience, run hot and cold water pipes parallel to one another, between 3" and 6" apart. Use the straightest, most direct routes possible when planning the layout, because too many bends in the pipe runs can cause significant friction and reduce water pressure.

It is a good idea to removed old supply pipes that are exposed, but pipes hidden in walls can be left in place unless they interfere with the installation of the new supply pipes.

How to Replace Water Supply Pipes

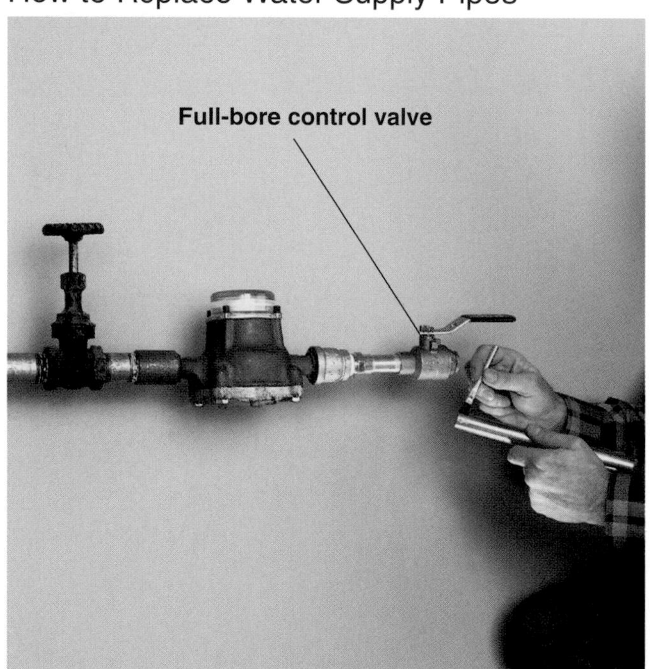

1 Shut off the water on the street side of the water meter, then disconnect and remove the old water pipes from the house side. Solder a ¾" male threaded adapter and full-bore control valve to a short length of ¾" copper pipe, then attach this assembly to the house side of the water meter. Extend the ¾" cold-water distribution pipe toward the nearest fixture, which is usually the water heater.

2 At the water heater, install a ¾" T-fitting in the cold-water distribution pipe. Use two lengths of ¾" copper pipe and a full-bore control valve to run a branch pipe to the water heater. From the outlet opening on the water heater, extend a ¾" hot-water distribution pipe, also with a full-bore control valve. Continue the hot and cold supply lines on parallel routes toward the next group of fixtures in your house.

3 Establish routes for branch supply lines by drilling holes into stud cavities. Install T-fittings, then begin the branch lines by installing brass control valves. Branch lines should be made with ¾" pipe if they are supplying more than one fixture; ½" if they are supplying only one fixture.

4 Extend the branch lines to the fixtures. In our project, we are running ¾" vertical branch lines up through the framed chase to the bathroom. Route pipes around obstacles, such as a main waste-vent stack, by using 45° and 90° elbows and short lengths of pipe.

5 Where branch lines run through studs or floor joists, drill holes or cut notches in the framing members (page 96), then insert the pipes. For long runs of pipe, you may need to join two or more shorter lengths of pipe, using couplings as you create the runs.

6 Install ¾" to ½" reducing T-fittings and elbows to extend the branch lines to individual fixtures. In our bathroom, we are installing hot and cold stub-outs for the bathtub and sink, and a cold-water stub-out for the toilet. Cap each stub-out until your work has been inspected and the wall surfaces have been completed.

INDEX

HANDYMAN
CLUB OF AMERICA